Armando S. Bevelacqua

Hazardous Materials
Managing the Incident | FIELD OPERATIONS GUIDE
SECOND EDITION

JONES & BARTLETT
LEARNING

World Headquarters
Jones & Bartlett Learning
5 Wall Street
Burlington, MA 01803
978-443-5000
info@jblearning.com
www.jblearning.com

Jones & Bartlett Learning books and products are available through most bookstores and online booksellers. To contact Jones & Bartlett Learning directly, call 800-832-0034, fax 978-443-8000, or visit our website, www.jblearning.com.

Substantial discounts on bulk quantities of Jones & Bartlett Learning publications are available to corporations, professional associations, and other qualified organizations. For details and specific discount information, contact the special sales department at Jones & Bartlett Learning via the above contact information or send an email to specialsales@jblearning.com.

Copyright © 2014 by Jones & Bartlett Learning, LLC, an Ascend Learning Company

All rights reserved. No part of the material protected by this copyright may be reproduced or utilized in any form, electronic or mechanical, including photocopying, recording, or by any information storage and retrieval system, without written permission from the copyright owner.

The content, statements, views, and opinions herein are the sole expression of the respective authors and not that of Jones & Bartlett Learning, LLC. Reference herein to any specific commercial product, process, or service by trade name, trademark, manufacturer, or otherwise does not constitute or imply its endorsement or recommendation by Jones & Bartlett Learning, LLC and such reference shall not be used for advertising or product endorsement purposes. All trademarks displayed are the trademarks of the parties noted herein. *Hazardous Materials: Managing the Incident, Field Operations Guide, Second Edition* is an independent publication and has not been authorized, sponsored, or otherwise approved by the owners of the trademarks or service marks referenced in this product.

There may be images in this book that feature models; these models do not necessarily endorse, represent, or participate in the activities represented in the images. Any screenshots in this product are for educational and instructive purposes only. Any individuals and scenarios featured in the case studies throughout this product may be real or fictitious, but are used for instructional purposes only.

All trademarks displayed are the trademarks of the parties noted therein. Such use of trademarks is not an endorsement by said parties of Jones & Bartlett Learning, its products, or its services, nor should such use be deemed an endorsement by Jones & Bartlett Learning of said third party's products or services.

Hazardous materials emergency response work is extremely dangerous and many emergency responders have died or sustained serious injury and illness while attempting to mitigate an incident. There is no possible way that this text can cover the full spectrum of problems and contingencies for dealing with every type of emergency incident. The user is warned to exercise all necessary cautions when dealing with hazardous materials. Always use a risk-based decision making process and place personal safety first.

It is the intent of the authors that this text be a part of the user's formal training in the management of hazardous materials emergencies. Even though this book is based on commonly used practices, references, laws, regulations, and consensus standards, it is not meant to set a standard of operations for any emergency response organization. The users are directed to develop their own written Standard Operating Procedures, which follow all system, agency, or employer guidelines for handling hazardous materials. It is the user's sole responsibility to stay up to date with procedures, regulations, and product developments that may improve personal health and safety.

Production Credits
Chief Executive Officer: Ty Field
President: James Homer
Chief Product Officer: Eduardo Moura
Executive Publisher: Kimberly Brophy
Executive Acquisitions Editor: William Larkin
Associate Managing Editor: Janet Morris
Production Editor: Cindie Bryan
Vice President of Sales, Public Safety Group: Matthew Maniscalco
Director of Sales, Public Safety Group: Patricia Einstein
Senior Marketing Manager: Brian Rooney
VP, Manufacturing and Inventory Control: Therese Connell
Composition: Aptara®, Inc.
Cover Design: Scott Moden
Director of Photo Research and Permissions: Amy Wrynn
Photo Research and Permissions: Dave Millar
Cover Image: © Courtesy of U.S. Coast Guard (top left) and
    Rob Schnepp (bottom left and right)
Printing and Binding: Courier Companies
Cover Printing: Courier Companies

ISBN: 978-1-4496-9672-6

6048

Printed in the United States of America
17 16 15 14 13   10 9 8 7 6 5 4 3 2 1

How to Use This Book

The hazardous materials (hazmat) incident has become more than simply an accidental event. Although it was always a possibility, the criminal or intentional use of hazardous substances has become more prevalent in recent years. This has required an increased knowledge of response and the dangerous work that an accident may involve. This text is designed to enhance the responder's ability to manage such an incident. It is divided into six areas of incident hierarchy.

SECTION 1: The Eight Step Process©

Section 1 is an overview of the Eight Step Process©, which outlines the linear approach toward managing the hazmat incident. Page references are provided to both this text and the main text, *Hazardous Materials: Managing the Incident, fourth edition*, where you will find additional information on certain topics addressed in this section.* This section also identifies the issues presented by a chemical, biological, radiological, nuclear, or explosive (CBRNE) incident.

SECTION 2: First Response Actions

Section 2 discusses the factors that first responders would need to consider and the functions they may need to perform before the hazmat team arrives, including recognition and identification issues that can arise during the initial phases of an incident.

SECTION 3: Strategic Goals

Section 3 is the Incident Commander section, addressing the overall strategic goals for hazmat response. These goals are organized according to the responsibilities of the safety officer, the public information officer, and the liaison officer. This section also covers the necessary terminal functions of the scene managment.

SECTION 4: Tactical Goals

Section 4 discusses the Hazmat Branch and the functions of the hazmat team. Each of this group's tasks is identified, along with an overall action plan and a tactical worksheet. Page references for further coverage are again provided to both this text and the fourth edition of *Hazardous Materials: Managing the Incident*.

SECTION 5: Top 35 Chemicals with CBRNE

Section 5 addresses the leading 35 chemicals of concern in a hazmat incident. These chemicals either have been found to cause injury or are significant due to their high frequency of release or quantities in which they are transported. The "scan sheets" list the makeup and toxicology of each chemical, as well as the appropriate personal protective equipment (PPE) and decontamination methods.

SECTION 6: CBRNE Tables

Section 6 contains tables of CBRNE materials, providing a quick reference to the general chemicals, chemical agents that serve as illicit drug precursors, and the biological, isotope, nuclear, and explosive descriptions.

---

*Note: Page references marked "4th edition" refer to the main text, *Hazardous Materials: Managing the Incident, fourth edition*. All other page references are to *Field Operations Guide, second edition*.

INTRODUCTION

Contents

Smoke: © Greg Henry/ShutterStock, Inc.; Flames: © Photos.com

SECTION 1 — The Eight Step Process© / 1

1. Site Management and Control 2
2. Identify the Problem 6
3. Hazard and Risk Evaluation 10
4. Select Personal Protective Clothing and Equipment 14
5. Information Management and Resource Coordintation 18
6. Implement Response Objectives 22
7. Decontamination 26
8. Terminating the Incident: Restoration and Recovery 30

SECTION 2 — First Response Actions / 35

First Responder Initial Actions 40
Placards 42
Transportation Overview: Damage Assessment 44
Cargo Tank Trucks/Containers 45
Railroad Tank Cars 51
Nonbulk Packaging 56
Atmospheric and Low-Pressure Liquid Storage Tanks 58
Railroad Tank Car Markings 61
CBRNE Signs and Symptoms Matrix: Chemical Matrix 72

SECTION 3 — Strategic Goals / 77

Incident Command 78
Public Information 80
Liaison Officer 82
Terminating Officer 84

SECTION 4 — Tactical Goals / 87

Hazmat Operations 90
Information 94
Recon/Entry 96
Hazmat Medical 101
Resources 105
Decontamination 110

SECTION 5 — Top 35 Chemicals with CBRNE / 115

Chemical Scan Sheets 116

SECTION 6 — CBRNE Tables / 165

General Chemical Descriptions 166
Chemical Agent Illicit Drug Precursors 167
Biological Descriptions 170
Isotope Descriptions 173
Nuclear Descriptions 175
Explosive Descriptions 177
Estimations for Volume Using Formulae 178

The Eight Step Process©

SECTION 1

1. Site Management and Control
2. Identify the Problem
3. Hazard Assessment and Risk Evaluation
4. Select Personal Protective Clothing and Equipment
5. Information Management and Resource Coordination
6. Implement Response Objectives
7. Decontamination
8. Terminating the Incident

1. SITE MANAGEMENT AND CONTROL

Approach and Position

❑ Approach from an uphill/upwind direction
- Contact incoming units of safe direction of travel
- If direction is not possible, establish maximum distance
  - Distances (page 40)

❑ Conditions found through observation (visible clues)
- Mode of transportation (container)
  - Road (page 45)
  - Rail (page 51)
  - Water international codes and intermodal (pages 63, 64, 69)
- Occupancy
  - Heavy industrial
  - Light industrial
  - Product transfer terminal
- Incident type (visible clues)
  - Spill
  - Leak
  - Release
  - Container

Site Management

❑ Establish command
- Avoid areas that can produce scene congestion

❑ Isolate the immediate area (1,000 ft for flammable/toxic releases)
- Ensure hazard control zones
- Ensure access control points

❑ Establish hazmat branch

❑ Request additional resources (see Staging Areas)
- Hazmat team
- Environmental protection
- United States Coast Guard
- Department of Transportation
- Nuclear regulation

❑ Contact state or local EOC

❑ Manage exposures
- Human and physical exposures (see Public Protection)

Dispatch Considerations

Weather conditions (page 14)
Location/occupancy
    Industrial process (pages 16, 22, 58)
    Fixed storage (page 19)
Preplans available
    Facilities papers (page 10)
    MSDS
Transportation mode

Roadway	DOT (pages 145, 180)
Railhead	DOT (pages 145, 180)
Waterway	USCG (pages 23, 180)

Threat Hazard

Target Analysis

Operational Conditions
Multiple indicators
Multiple calls for injured
Multiple calls for specified illness
Bomb call with/without time frame
Avoid choke points
Designate rally points

Risk Assessment
Human and physical exposures
Placement within the community
    Roadway
    Railhead
    Waterway (Seaport)
    Airport
Conditions presented (solid, liquid, gas)

Probability Assessment
Multiple intelligence sources
Significant indications in which threat condition established
Current threat condition

■ Staging Areas
- ❏ Identify staging area
  - Identify the immediate hazard area
  - Identify specific locations for additional resources
- ❏ Assign an officer for Level II staging maintenance
  - Fire apparatus
  - Law enforcement
  - Contractors
  - Heavy equipment
  - Support operations
    - Fuel truck
    - Water
    - Facilities
    - Food

■ Public Protection
- ❏ Hazard condition
  - Uncontrollable chemical fire
  - Airborne chemical release
  - Population load close to the hazard
- ❏ Evacuation
  - Remove all persons from the immediate area
  - DOT-ERG with plume models for immediate area established
  - Relocation facility
- ❏ Protect in place (type of predominate structures may necessitate evacuation)
  - Refuge area defined
  - Buildings, windows, and doors are shut
  - HVAC, heating, and air conditioning are turned off
  - Have population monitor a local radio/TV station
- ❏ Public information
  - Relocation points
  - Public alert and notification
  - Detailed information
    - News briefs (set time during incident)

Security of the Incident
- Liaison at the command post with law enforcement
- Establish isolation perimeter with control points
- Enforcement of secured areas
  - Staging:
    - Fire apparatus
    - Contractors
    - Heavy equipment
    - Support operations
- Consider all incidents to be crime scenes until proven otherwise
- If incident is intentional
  - Law enforcement needs for entry
  - Evidence, crime scene investigation
  - Public protection actions (direction)
- Potential immediate notifications
  - Law enforcement (city, county, state)
  - Health department
  - Hospitals
  - Emergency operations center(s)
  - Public works
  - Media
  - Contingency contractors

THE EIGHT STEP PROCESS©/1. Site Management and Control

Interrelated Strategic and Tactical Objectives

Smoke: © Greg Henry/ShutterStock, Inc.; Flames: © Photos.com

1. SITE MANAGEMENT AND CONTROL

GOAL: To establish the playing field so that all subsequent response operations can be implemented both safety and effectively.

FUNCTION: Site management and control consists of managing and securing the physical layout of the incident. You cannot safely and effectively manage the incident if you do not have control of the scene. As a result, site management and control is a critical benchmark in the overall success of the response and is the foundation upon which all subsequent response functions and tactics are built.

■ Approach and Position

❑ Approach from an uphill/upwind direction
- Contact incoming units
- Establish maximum distances

❑ Conditions found through observation (visible clues)
- Transportation
- Occupancy
- Incident type

■ Site Management

❑ Establish command
- Avoid scene congestion

❑ Isolate immediate area (1,000 ft for flammable/toxic releases)
- Ensure hazard control zones and access control points

❑ Establish hazmat branch

❑ Request additional resources (see Staging Areas)

❑ State warning point

❑ Exposures
- Identify human and physical exposures

		4th edition
Survey the Incident	page 6	100
Defining Criteria	page 6	148–171
Behavioral Event	page 11	101
Tactical Planning	page 18	7, 19, 321
Chemical Safety Sheets	page 116	

		4th edition
Identification	page 7	100
Hazard Assessment	page 10	55, 100
Hazard/Risk Management (Rescue/Public Protection)	page 10	323
PPE Compatibility	page 14	55

■ Staging Areas

- ❏ Identify staging areas
- ❏ Assign an officer for Level II staging maintenance
    - Supportive equipment
        - Fire apparatus
        - Law enforcement
        - Contractors
    - Support
        - Fuel trucks
        - Water
        - Facilities
        - Supplies

■ Public Protection

- ❏ Hazard condition
    - Uncontrollable conditions
        - Airborne
        - Population load
- ❏ Evacuation
    - Plume models
    - DOT-ERG
    - Relocation facilities
- ❏ Protect in place
    - Refuge area defined
    - Building shut with HVAC turned off
    - Population to monitor media
- ❏ Public information
    - Relocation points
    - Public alert
    - Detailed media release

		4th edition
Survey the Incident	page 6	100
Defining Criteria	page 6	148–171
Resource Coordination	page 19	308

		4th edition
Survey the Incident	page 6	100
Defining Criteria	page 6	148–171
Behavior Event	page 11	101, 237
Hazard/Risk Assessment	page 10	55, 100

Threat Hazard Relationships

- Analysis of operational conditions as they are correlated with incident intelligence
- Development of an incident intelligence plan with law enforcement
- Evaluate potential risks
    CBRNE used and its impact
    Perpetrator resistance
    Perimeter control as it relates to resistance
    Force protection as it relates to resistance
- Strong public protection plan based upon the CBRNE/human conditions
- Strong incident control (i.e., hazard control zones, access checkpoints, and areas of refuge)

THE EIGHT STEP PROCESS©/1. Site Management and Control

2. IDENTIFY THE PROBLEM

Survey the Incident

❏ Surrounding conditions
- Weather parameters
- Endangered populations
- Topography
- Exposures

❏ Hazard behavior
- Smoke/clouds/plumes
- Spill/leak/rupture

Defining Criteria

❏ Occupancy and location
- Type of manufacture process
- Chemical exposures

❏ Container shapes and sizes (pages 45, 63)
- Size of container(s)
- Condition of the container(s)
- Type of container(s)

❏ Markings and colors (pages 60–62, 90)
- Color codes
- Container specification number
- Signal words
- Contents name

❏ Labels and placards (pages 40, 42)
- Background color
- Hazard class symbol
- Hazard class/division number
- Four-digit ID number

❏ Shipping papers and facility documents (page 65)
- Bill of lading
- Waybill consist
- Dangerous cargo manifest
- Air bill
- Material Safety Data Sheet

❏ Monitoring and detection equipment
- Chemical family or classification

Influencing Factors (Based on potential enormity)

Defensive: Reconnaissance
Operational objective: To obtain information on site layout, physical hazards, access, and related conditions.
Operational action: Obtained through preplans, human observation, and physical conditions of the site.

Offensive: Reconnaissance
Operational objective: To obtain incident conditions, *which cannot be observed from a defensive position.*
Operational action: Monitoring, sampling, damage assessment, offensive reconnaissance, two in—two out. May be combined with offensive control operations.

Threat Hazard

Target Analysis
Operational conditions
- Potential quantity of material
- Potential exposure
  - Chemical/exposure
- Future weather conditions
  - Temperature
  - Wind direction/speed
  - Humidity

Risk Assessment
- Weapons involved
- Temperature inversions
- Medical implications

Probability Assessment
- Feasible potential
- Behavioral inclinations
- Operational resolve

- Senses
  - Observational clues

Identification
- ❏ Type of chemical involved
- ❏ Chemical and physical properties
  - Ambient temperature
  - Size of release (surface area)
  - Vapor pressure
  - Solid/liquid
    - Melting point/boiling point
    - Specific gravity
    - Water solubility
    - Vapor density
  - Gas—flammability/toxicity
    - Flammable range
    - Ignition temperature
    - Boiling point
  - Reactivity
    - Air reactive
    - Water reactive
    - Hypergolic
    - Polymerization
    - Incompatibilities
  - Information evaluation
    - Match visible clues with potential products

Classification
- ❏ Behavioral event (page 11)
- ❏ Damage assessment (page 10)
- ❏ Monitoring and detection
  - Chemical analysis
    - Chemical classification
    - Systematic analysis

Verification
- ❏ Evaluate and correlate classification perimeters
  - Chemical and physical properties
    - Monitoring/detection technologies correlate
  - Seven defining criteria
- ❏ Verify conditions and factors as presented

Future Considerations
- Weather conditions
- Resource potential
  - Working crews
  - Resting crews
  - Support staff
  - Contractors
  - Specialized staff
- Population effected
  - Public protective actions

THE EIGHT STEP PROCESS© / 2. Identify the Problem

Interrelated Strategic and Tactical Objectives

Smoke: © Greg Henry/ShutterStock, Inc.; Flames: © Photos.com

2. IDENTIFY THE PROBLEM

GOAL: To identify the scope and nature of the problem, including the type and nature of hazardous materials involved as appropriate.

FUNCTION: Identify the scope and nature of the problem. This includes recognition, identification, and verification of the hazardous materials involved in the incident; type of container, as appropriate; and exposures.

Methods of identification include analyzing container shapes, markings, labels and placards, and facility documents (e.g., Material Safety Data Sheets [MSDS]); using monitoring and detection equipment; and identifying by the senses (i.e., physical observations, smell, reports from victims). Responders should remember that even when the hazardous substances involved have been identified, the information should always be verified.

■ Survey the Incident

❏ Surrounding conditions
- Endangered populations
- Weather
- Topography

❏ Hazard behavior
- State of matter
- Containment system

■ Defining Criteria

❏ Occupancy and location
- Type of manufacturing process

❏ Container shapes and sizes
- Size of containers
- Condition of containers

❏ Markings and colors
- Codes and specifications

❏ Labels and placards

❏ Shipping papers and facility documents

❏ Monitoring and detection equipment

		4th edition
Approach and Position	page 2	115
Site Management	page 2	114
Risk Assessment	page 11	142

		4th edition
Exposure Assessment	page 10	33
Tasks to Be Performed	page 78	231
Information Management	page 18	308
Tactical Objectives	page 23	324
Decontamination	page 26	395
Site Selection	page 26	407

8

■ Identification

- ❑ Type of chemical
- ❑ Chemical and physical properties
  - ■ Ambient temperature
  - ■ Size of release
  - ■ Vapor pressure
  - ■ Solid/liquid
  - ■ Gas
  - ■ Reactivity
  - ■ Information evaluation

■ Classification

- ❑ Behavioral event
- ❑ Damage assessment
- ❑ Monitoring and detection
  - ■ Chemical classification

■ Verification

- ❑ Evaluate and correlate classification
  - ■ Chemical and physical properties
  - ■ Seven defining criteria
- ❑ Verify conditions and factors presented
  - ■ Correlate found data to the incident

		4th edition
Hazard Assessment	page 10	55, 100
Exposure Assessment	page 10	33
Risk Assessment	page 11	142

		4th edition
Passive Analysis	page 10	148–171
Active Analysis	page 10	206
Information Management	page 18	308

		4th edition
Hazard and Risk Evaluation	page 10	100
Tactical Planning	page 78	324
Resource Coordination	page 18	308

Threat Hazard Relationships

- ■ Sweep the entire operational area for secondary and tertiary devices
- ■ Through the incident intelligence plan, identify the probability of the incident
  - Feasible potential
  - Behavioral inclinations
  - Operational resolve
- ■ Classify event and additional ancillary incidents within the response system

THE EIGHT STEP PROCESS©/2. Identify the Problem

3. HAZARD AND RISK EVALUATION

Hazard Assessment

❑ Identify the hazard
- Survey the incident (page 2)
- Identify defining criteria (page 6)
- Identification process has been completed (page 7)

❑ Behavioral event (risk assessment)

❑ Damage assessment (risk assessment)

Exposure Assessment

❑ Reference materials
- Hazard data and information

❑ Passive analysis
- Observation
- Referencing of defining criteria with hazard data
  - Corrosivity
  - Flammability
  - Oxidizing potential
  - Oxygen potential
  - Radioactivity
  - Toxicity

❑ Active analysis
- Reconnaissance
  - Relate to defining criteria with hazard data
  - Correlate with chemical and physical properties
- Monitoring and detection
  - Corrosivity
    pH paper, strips, meters
  - Flammability
    CGI
  - Oxygen potential
    Oxygen peter
  - Radioactivity
    Radiation detector, dosimeters
  - Toxicity
    Colorimetrics

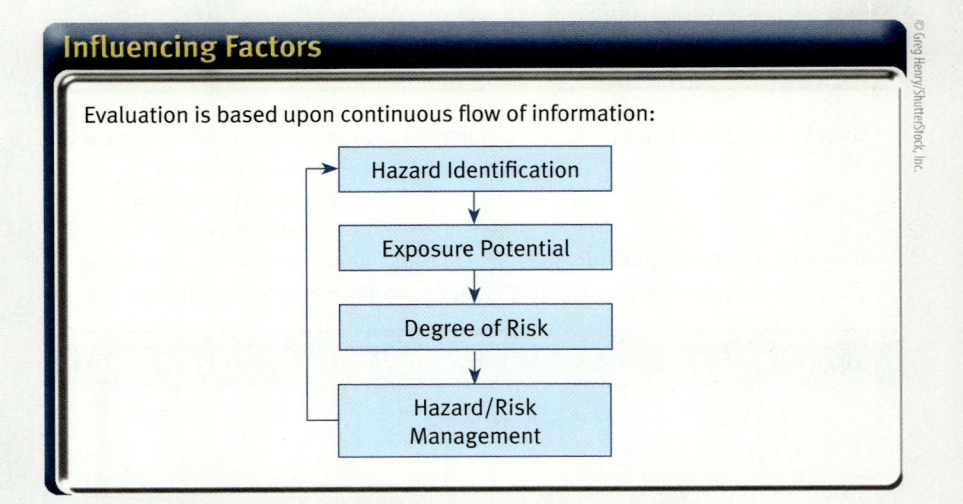

Influencing Factors

Evaluation is based upon continuous flow of information:

Hazard Identification → Exposure Potential → Degree of Risk → Hazard/Risk Management

Threat Hazard

Target Analysis

Operational Conditions
Quantity of material(s)
Ease of perpetration
Time of day
Significant event(s)

Risk Assessment
Population affected
Toxicity of material(s)
Chemical and physical properties
Weaponry

Probability Assessment
Behavior of material(s) correlated to the event in progress
Behavior of the material(s) correlated to potential events
(Behavior events)

Risk Assessment
- ❏ Quantity of material
  - Potential additional releases
  - Dispersion characteristics
- ❏ Containment systems (pages 44–59)
  - Correlate to defining criteria
  - Stress event (behavior)
  - Physical damage (damage potentials)
    - Behavior event (stress)
      - Breach
      - Release
      - Engulf
      - Impingement
      - Harm
    - Damage potentials of containers
      - Crack
      - Score
      - Gouge
      - Wheel burn
      - Dent
      - Rail burn
      - Street burn

Hazard/Risk Management Implementing Response Objectives
- ❏ Step 6 (page 22)
  - Rescue
  - Public protective actions
  - Spill control
  - Leak control
  - Fire control
  - Recovery

Damage Potentials

- Dispersion characteristics: based on the stresses—thermal, mechanical, chemical, irradiation, and etiological, or combination thereof
- Other factors considered:
  Ambient temperature corrected to the material(s) involved
  Weather condition present and future topography (urban, rural)
- Stress events (behavior event):

Breach	Disintegration
	Runaway cracking
	Failure of container attachments
	Container punctures, splits, or tears
Release	Detonation
	Violent rupture
	Rapid relief
	Spills or leaks
Engulfing	Cloud/plume production
	Following the path of least resistance
Impingement	Time dependent
	Chemical and quantity dependent
	Form of threat (solid, liquid, gas, radiological, etiological; see Harm)
Harm	Thermal
	Radiation
	Asphyxiation
	Toxicity
	Corrosivity
	Etiological
	Mechanical

THE EIGHT STEP PROCESS©/3. Hazard and Risk Evaluation

Interrelated Strategic and Tactical Objectives

Smoke: © Greg Henry/ShutterStock, Inc.; Flames: © Photos.com

3. HAZARD AND RISK EVALUATION

GOAL: To assess the hazards present, evaluate the level of risk, and establish an incident action plan (IAP) to make the problem go away.

FUNCTION: This is THE most critical function that public safety personnel perform. The primary objective of the risk evaluation process is to determine whether responders should intervene and which strategical objectives and tactical options should be pursued to control the problem at hand. You can't get this wrong. If you lack the expertise to perform this function adequately, get help from someone who can provide that assistance, such as local HMRTs and product/container specialists.

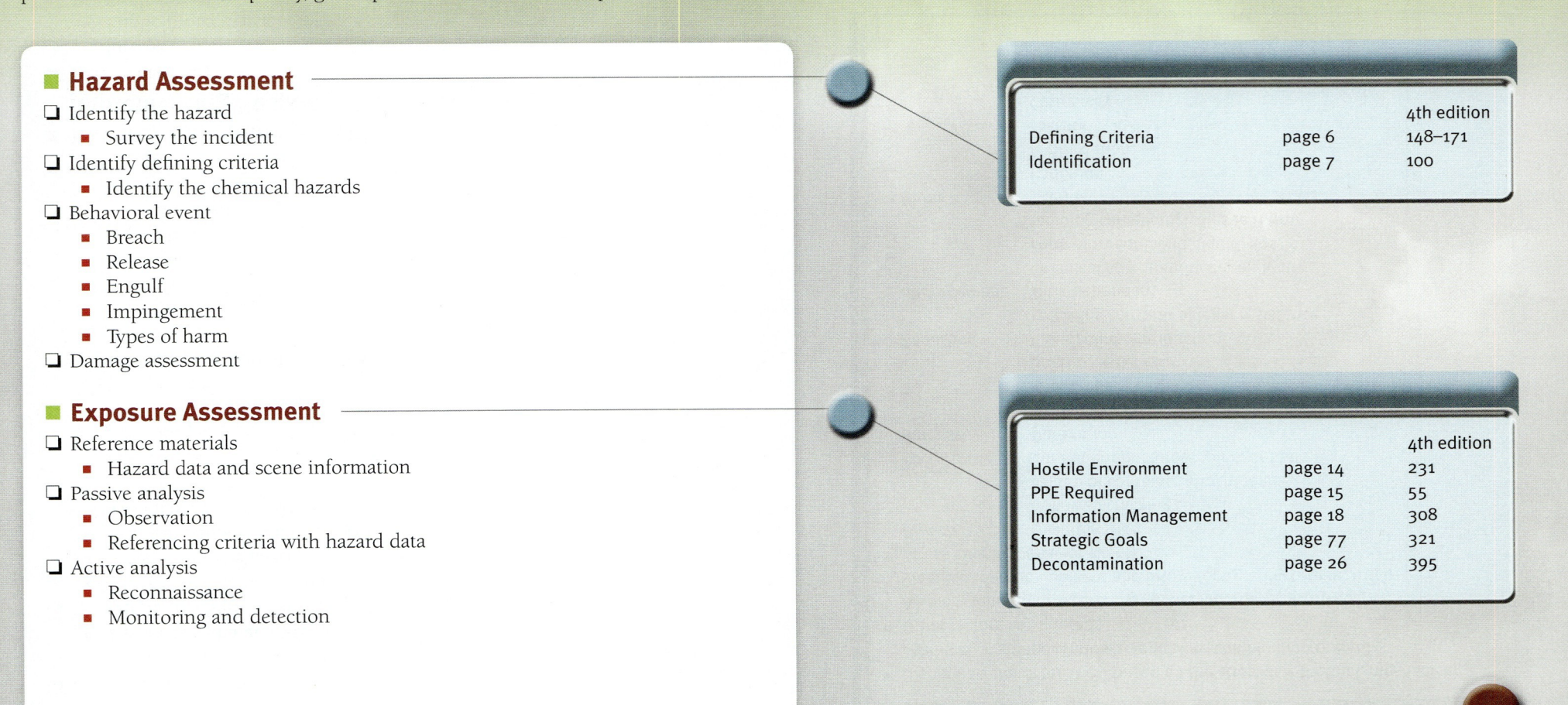

■ Hazard Assessment

❑ Identify the hazard
  ■ Survey the incident
❑ Identify defining criteria
  ■ Identify the chemical hazards
❑ Behavioral event
  ■ Breach
  ■ Release
  ■ Engulf
  ■ Impingement
  ■ Types of harm
❑ Damage assessment

		4th edition
Defining Criteria	page 6	148–171
Identification	page 7	100

■ Exposure Assessment

❑ Reference materials
  ■ Hazard data and scene information
❑ Passive analysis
  ■ Observation
  ■ Referencing criteria with hazard data
❑ Active analysis
  ■ Reconnaissance
  ■ Monitoring and detection

		4th edition
Hostile Environment	page 14	231
PPE Required	page 15	55
Information Management	page 18	308
Strategic Goals	page 77	321
Decontamination	page 26	395

■ Risk Assessment
- ❏ Quantity of material
  - ▪ Additional potential releases
  - ▪ Dispersion characteristics
- ❏ Containment systems
  - ▪ Crack
  - ▪ Score
  - ▪ Gouge
  - ▪ Wheel burn
  - ▪ Dent
  - ▪ Rail burn
  - ▪ Street burn

■ Hazard/Risk Management
Implementing Response Objectives
- ❏ Rescue
- ❏ Public protective actions
- ❏ Spill control
- ❏ Leak control
- ❏ Fire control
- ❏ Recovery

		4th edition
Event Size-up	page 10	242
Tactical Objectives	page 40	324

		4th edition
Strategic Goals	page 77	321
Tactical Objectives	page 40	324
Decontamination	page 26	395

Threat Hazard Relationships

- Evaluate the operational intent in conjunction with the behavior event and damage assessment
- Ensure that the level of PPE for all entry teams is appropriate based upon the intelligence gathered on type of the material, referenced hazards, and the type of technical environment (page 14)
- Discuss with all involved agencies:
    - Operational conditions
    - Risk assessment
    - Probability assessment
    - Threat factors steps 1 and 2 (pages 2, 6)

THE EIGHT STEP PROCESS©/3. Hazard and Risk Evaluation

4. SELECT PERSONAL PROTECTIVE CLOTHING AND EQUIPMENT

Hostile Environment

❑ Degree of hazard
- Action plan developed (page 78)
  - Type of mission versus degree of hazard
    Identify the response objectives (page 22)
  - Detail possible chemical/physical outcomes based
    Referenced material
    Environmental concerns
    Resource capability

❑ Potential outcomes
- Dispersion patterns
- Size, shape, and concentrations associated with release
- Correlate with defined criteria and potential behavior of the container

❑ Compatibility of PPE with the chemical(s)
- Identify the suit compatibility
- Match suit and glove compatibility
- Identify the level of respiratory protection

Tasks to Be Performed

❑ Entry and backup mission
- Hot zone mission objectives identified
- Based on suit limitations/advantages
- Objectives correlated with mission and availability of suits

❑ Decontamination team mission level
- Warm zone mission objectives identified
- Based on suit limitations/advantages
- Objectives correlated with mission and availability of suits

❑ Area of refuge mission
- Warm zone safety mission identified

❑ Support teams
- Equipment
- Medical

Factors

Type of mission versus degree of hazard

- Exposures/harm
  Human exposures (rescue potential)
  Lethality of chemical (nonrescue)
  Property damage potential
  Environmental potential
- Scene orientation
  Defensive
  Offensive
  Nonintervention
- Identify the degree of potential harm
  Dependent on level of engagement (work area) versus possible level of protection
  Entry and backup team mission level

Compatibility

- Degradation—temperature dependent measured at 70 –75°F
- Penetration—suit testing and inspection schedules
- Permeation—breakthrough time (Most charts describe this as breakthrough time and permeation rates.)

Threat Hazard

Target Analysis

Operational Conditions
Base suit selection on chemical conditions versus mission objectives

Risk Assessment
Objectives of the mission must take into consideration external threat influences

Probability Assessment
Consider external influences such as changing threat conditions, chemical versus SWAT objectives

PPE Required for the Mission
- Identify the factors considered for suit selection
  - Chemical versus available suit compatibility
  - Duration of operation
  - Degree of decontamination required
  - Availability of human resources
- Safety of operational objectives
  - Monitoring of personnel
  - Backup personnel
  - Hand signals and communications
  - Medical considerations
    - Heat stress
    - Rehabilitation
- Approval of suit/respiratory protection from incident commander

Capabilities of the User
- Identify operational outcomes
  - Mission objectives for each team
  - Use of communications
  - Safety of mission
    - Based upon environmental concerns versus mission options
    - Human resources available
- Safety of operational objectives
  - Medical monitoring of entry/backup personnel
    - Medical considerations
    - Heat stress
    - Rehabilitation
  - Hand signals and communications
  - Emergency entry procedures
    - Backup team advised of operational objectives
    - Alternate response objectives identified
      - Rescue options
      - Initial scene stabilization

Personnel Protection Equipment Selection

PPE selection is based on:
- Scene characteristics
  - Chemistry of the material
  - Behavior of the container
  - Environmental influences
- Life safety
  - Safety of the responders
  - Safety of the population
- Incident stabilization (effects on large populations)
  - Degree of information readily available
  - Resources able to accomplish aggressive goals
- Protective actions
  - Area of assignment
  - Mission objectives
    - Hazard conditions
    - Evacuation
    - Protect in place
    - Public Information

THE EIGHT STEP PROCESS©/4. Select Personal Protective Clothing and Equipment

Interrelated Strategic and Tactical Objectives

Smoke: © Greg Henry/ShutterStock, Inc.; Flames: © Photos.com

4. SELECT PERSONAL PROTECTIVE CLOTHING AND EQUIPMENT

GOAL: To ensure that all emergency response personnel have the appropriate level of personal protective clothing and equipment for the expected tasks.

FUNCTION: Based upon the results of the hazard and risk assessment process, emergency response personnel will select the proper level of personal protective clothing and equipment. Two primary types of personal protective clothing are commonly used at hazmat incidents: (1) structural firefighting protective clothing and (2) chemical-protective clothing.

■ Hostile Environment

❏ Degree of hazard
- Action plan developed
- Type of mission

❏ Potential outcomes
- Dispersion patterns
- Size and concentrations of release
- Correlate with defining criteria
- Correlate potential behavior of the container

❏ Compatibility of PPE with the chemical(s)
- Hazard risk assessment
- Correlate with protection on hand

		4th edition
Survey the Incident	page 6	100
Defining Criteria	page 6	148–171
Identification	page 7	100
Risk Assessment	page 11	55, 100, 243

■ Tasks to Be Performed

❏ Entry and backup
❏ Decontamination
❏ Area of refuge
❏ Support teams

Hazard and Risk		4th edition
Evaluation	page 11	33, 243
Strategic Goals	page 77	321
Tactical Objectives	page 40	324

PPE Required for the Mission

- ❏ Identify the factors for suit selection
    - Available suit compatibility
    - Duration of the operation
    - Degree of decontamination required
    - Available human resources
- ❏ Safety factors for operational objectives
    - Monitoring of personnel
    - Available backup personnel
    - Hand signals and communication
    - Medical considerations
- ❏ Approval from the commander

Capabilities of the User

- ❏ Identify operational outcomes
    - Mission objectives for each entry
    - Communication
    - Potential safety issues
- ❏ Safety of operational objectives
    - Medical monitoring of entry and backup
    - Hand signals and communication channels
    - Emergency communication
- ❏ Emergency entry procedures
    - Backup team operational objectives
    - Alternative response objectives defined

		4th edition
Tactical Objectives	page 40	33, 324, 243
Decontamination	page 26	395

		4th edition
Tactical Objectives	page 40	324
Decontamination	page 26	395

Threat Hazard Relationships

- Evaluate the type of hazardous environment
    - Degree of chemical hazard with perpetrator intent
    - Appropriately designed equipment for the mission's objective
- Ensure the level of PPE for all entry teams
    - Relates to the tasks or mission of the entry
    - Team capabilities match with tactical objective(s)
- Discuss and plan for a changing environment, due to:
    - CBRNE influences
    - Potential damage
    - Atmospheric changes
    - Multiple incidents within the system

5. INFORMATION MANAGEMENT AND RESOURCE COORDINATION

■ Information Management

❏ Strategic planning (Incident Command)
- Managing information
  - Incident command structure
  - Information distribution
- Interpretation of information
  - Information section/Incident command
  - Information gained
    - Hazards
    - PPE
    - Health considerations
    - Tactical recommendations
    - Decontamination
  - Formulate safety site plan
❏ Tactical Planning (Information Section) (pages 10, 18)
- Prioritization of information
  - Type of information
  - Data
    - Pre-Incident plans/familiarization
    - Emergency response plans
    - Pre-incident plans
  - Facts / Observations
    - Reference material
    - Visual clues
    - Defined criteria
    - Behavior events
  - First operational hour
    - Tactical recommendations
  - Long Range Planning
    - Strategic Recommendations
    - Contingency plan
- Information retrieval system (Paper/electronic)
  - Tactical worksheets
  - Memos between working groups (runners)
  - Radio communication

Factors

- Information Distribution: (page 39)
  - Information
  - Reconnaissance/entry
  - Resources
  - Hazmat medical
  - Decontamination
- Informational needs (short/long term)
  - Data/facts (pages 78, 90)
  - Preincident plan
    - Process flow diagrams
    - High consequence facility
    - Sensitive exposures
    - Water supply
    - High commodity load
    - Resource availability
    - Location restrictions
    - Access routes

Threat Hazard

Target Analysis
Operational Conditions
- Location restrictions
- Protection of access routes

Risk Assessment
- Limited access routes
- Dead-end routes
- Significant instillation/facility
- High commodity load
- High-consequence facility

Probability Assessment
- Facility dependency for the infrastructure
- High economic base region

Resource Coordination

- ❏ Internal (agency Incident Management System)
  - Implement Incident Management System (pages 38, 78)
  - Resource section manages resources
    - Special resource requirements placed in staging
    - Tracking of all available personnel
      - Coordinates with Staging section
      - On large incidents, coordinates with Logistics
    - Tracking of all available equipment
      - Coordinates with Staging section
      - Supply inventory coordination
        - Protective equipment
        - Spill and leak control equipment
        - Decontamination materials
        - Specialize tools
- ❏ External (unified command)
  - Identify personnel to communicate action plans
    - Determine meeting time frames
  - Determine activity of resource
    - Identify operational time period
    - Apply safety plan to the specific resource
  - Describe incident visually with micro/macro diagrams
  - Create organizational/planning flowchart
    - Listing of organization
    - Resources provided
    - Phone number (cell phone), radio designation
    - Assignment
  - Assign external resource to a section, branch, or sector
    - Assign to a responsible functional group
  - Information transfer from unified command to PIO
    - Political officials
    - Media (public protective actions) (page 3)

Public Agencies

Federal:
United States Coast Guard—National Strike Team
United States Environmental Protection Agency—Environmental Response Team
Federal Bureau of Investigation—Coordinator, Hazardous Materials Response Unit
Department of Energy—Nuclear Emergency Search Team
Centers for Disease Control and Prevention
Disaster Medical Assistance Team
United States Army Tech Escort Unit
Chemical/Biological Incident Response Team
Civil Support Teams
Bureau of Alcohol, Tobacco and Firearms

Local/State:
Mutual aid fire and law enforcement resources
Environmental Protection Agency
Public Health Department
Public works
Statewide mutual aid response plan

Interrelated Strategic and Tactical Objectives

Smoke: © Greg Henry/ShutterStock, Inc.; Flames: © Photos.com

5. INFORMATION MANAGEMENT AND RESOURCE COORDINATION

GOAL: To provide for the timely and effective management, coordination, and dissemination of all pertinent data, information, and resources between all of the players.

FUNCTION: Refers to proper management, coordination, and dissemination of all pertinent data and information within the ICS in effect at the scene. In simple terms, this function cannot be effectively accomplished unless a unified ICS organization is in place. Of particular importance is the ability to determine which incident factors are involved, which functions of the Eight-Step Process© have been completed, which additional information must be obtained, and which incident factors remain unknown.

■ Information Management

❏ Strategic planning
- Managing information
- Incident command structure that promotes information distribution
- Interpretation of information
  - Hazards
  - PPE
  - Health considerations
  - Tactical recommendations
  - Decontamination
❏ Tactical planning
- Prioritization of information
- Information retrieval systems
- Tactical worksheets
- Radio communications
- Memos runners

		4th edition
Threat Hazards Steps 1–7	pages 2, 6, 10, 14, 18, 22, 26, 30	
Defining Criteria	page 6	148–171
Identification	page 7	100
Hazard Assessment	page 10	55, 100
Exposure Assessment	page 11	33

Resource Coordination

- ❏ Internal
    - Manage resources internally
    - Tracking of personnel
    - Tracking of equipment
    - Special resources
- ❏ External
    - Identify personnel to communicate action plans
    - Determine activity or resource needed
    - Describe the incident
        - Description of the micro problems
        - Description of the macro environment
    - Create organizational flowchart
    - Assign resource
    - Information to the PIO (page 80)

		4th edition
Site Management	page 2	305
Staging Area	page 3	115
Strategic Objectives	page 78	321
Tactical Objectives	page 40	33, 243, 324
PPE Required for the Mission	page 14	55, 100, 243

Threat Hazard Relationships

- Ensure the tactical plan lends itself to achieving the identified strategic goals
    - Consider tactical and strategic conditions
    - Ensure the development of an intelligence plan
- Through unified command (page 78)
    - Develop a strategic plan
    - Identify jurisdictional limitations and restrictions
    - Discuss and plan for a changing environment, due to:
        - CBRNE influences
        - Potential damage
        - Atmospheric changes
        - Multiple incidents within the system
    - Correlate information with intelligence gathered
    - Develop a strong interagency resource allocation plan
    - Develop a command meeting between strategic and tactical personnel
        - Develop an incident action plan that addresses multijurisdictional/multiagency goals
        - Consolidated command structure to achieve the desired objective

THE EIGHT STEP PROCESS©/5. Information Management and Resource Coordination

6. IMPLEMENT RESPONSE OBJECTIVES

Event Size-up

- ❏ Events (past, present, future)
- ❏ Behavior event (page 11)
  - Past event profile
  - Present event profile
    - Event planning
      - Magnitude
      - Occurrence
      - Timing
      - Effects
      - Location
    - Risk assessment
  - Future event profile
    - Options for positive influence

Strategic Goals: Basis for Operational Strategy

- ❏ Behavior event prediction
  - Magnitude
  - Occurrence
  - Timing
  - Effects
  - Location
- ❏ Goals
  - Offensive mode
    - Rescue
    - Public protective actions
    - Spill control (confinement)
    - Leak control (containment)
    - Fire control
    - Recovery
  - Defensive mode
    - Public protective actions
    - Spill control (limited methods)
  - Nonintervention mode
    - Public protective actions
    - Resource management (page 18)

Size Up Evaluation

Magnitude	Quantity of hazardous material
	Type of stress
	Breach
	Release
	Engulf
	Impinge
Occurrence	Exposed materials
	Incompatibilities
Timing	Rate of chemical reaction (chemical properties)
	Product transfer
Effects	Redirect stress
	Shield stress
	Move stressed system
Location	Degree of harm based on:
	Geography
	Weather conditions
	Quantity of material
	Level of available resources

Threat Hazard

Target Analysis

Operational Conditions
- Secondary devices screened for
- Secured isolation perimeter
- Incident supplies route secured

Risk Assessment
- Quantity and properties of material
  - Level of harm
- Geographical assessment
  - Responders versus public protected
- Police tactical engagement

Probability Assessment
- Signature profile of the event
- Consult with Intelligence sources

■ Tactical Objectives: Basis for Tactical Decisions

- ❏ Rescue
  - Risk assessment (page 11)
  - Event profile (pages 7, 10)
- ❏ Public protective actions
  - Isolation and control
  - Evacuation
  - Protect in place
- ❏ Spill control (confinement)
  - Absorption
  - Adsorption
  - Covering
  - Damming
  - Diking
  - Dilution
  - Diversion
  - Dispersion
  - Retention
  - Vapor dispersion
  - Vapor suppression
- ❏ Leak control (containment)
  - Neutralization
  - Overpacking
  - Patching and plugging
  - Pressure isolation and reduction
  - Solidification
  - Vacuuming
- ❏ Fire control
  - Foam application
  - Other extinguishing application
  - Exposure protection
- ❏ Recovery
  - Product removal and transfer
  - Termination (Step 8, 30, 84)

Objective Considerations

	Search/relocation/area of refuge established
	Executing technical rescue
	Perimeter security
	Quantity of absorbent materials versus quantity of released material
	Incompatibility with adsorbents
	Temporary cover available and compatible
Damming	Specific gravity (SG)
	Overflow dam: SG > 1
	Underflow dam: SG < 1
Diking	Limitations to a temporary measure
	No available soil; area is concrete or asphalt
	Frozen ground
	Equipment availability
	Human resources available
Dilution	Criteria met before action is taken as a last resort
	Run-off and environmental concerns
	Not water reactive
	Will not generate a toxic gas
	Will not form a solid or precipitate
	Water solubility factors are high
Diversion	Placed ahead of the spill
	Consider speed and quantity of material
	The greater the speed/quantity, the greater the length and angle of the barrier
	Area of involvement increases
	Dispersion increases involved area
Retention	Must be employed with diversion or diking
	Requires resources
Vapor dispersion/ vapor suppression	Run-off and environmental concerns
	May create pockets of chemical concern
Neutralization	Quantity needed calculated and available
Overpacking	Type and quantity available
Patching and plugging	Dependent on the size of opening
Pressure isolation/ reduction	Referenced and equipment available—venting, flaring, hot tap, vent and burn
Solidification	Quantity of material versus available containment
Vacuuming	Dependent on quantity, venting secured

THE EIGHT STEP PROCESS©/6. Implement Response Objectives

Interrelated Strategic and Tactical Objectives

Smoke: © Greg Henry/ShutterStock, Inc.; Flames: © Photos.com

6. IMPLEMENT RESPONSE OBJECTIVES

GOAL: To ensure that the incident priorities (i.e., rescue, incident stabilization, environmental and property protection) are accomplished in a safe, timely, and effective manner.

FUNCTION: The phase in which responders implement the best available strategic goals and tactical objectives, which will produce the most favorable outcome. If the incident is in the emergency phase, this is when we "make the problem go away." Common strategies to protect people and stabilize the problem include rescue, public protective actions, spill control, leak control, fire control, and recovery operations. In simple terms, these strategies are typically implemented by fire and rescue units, with law enforcement responsible for all security and criminal-related issues.

If the incident is in the post-emergency response phase, the focus of response personnel will likely become scene safety, clean-up, evidence preservation (as appropriate), and incident investigation. Specific tasks will include (1) initial site entry and monitoring to determine the extent of the hazards present; (2) an evaluation of the scene to locate evidence that can be used to reconstruct the events leading up to the incident; (3) identification of the contributing factors that caused the incident; (4) interviewing of on-scene personnel and witnesses to corroborate the information obtained and opinions formed based on the available data; and (5) documentation of preliminary results.

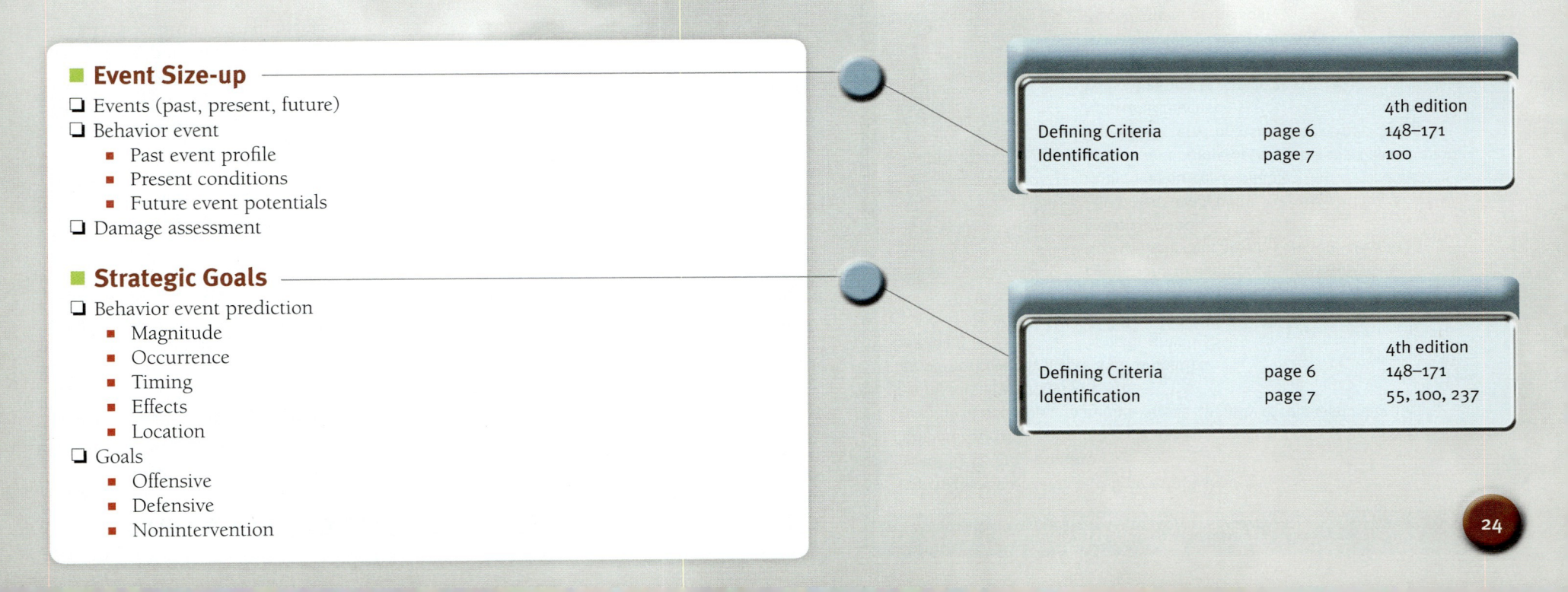

■ Event Size-up
- ❏ Events (past, present, future)
- ❏ Behavior event
  - ▪ Past event profile
  - ▪ Present conditions
  - ▪ Future event potentials
- ❏ Damage assessment

		4th edition
Defining Criteria	page 6	148–171
Identification	page 7	100

■ Strategic Goals
- ❏ Behavior event prediction
  - ▪ Magnitude
  - ▪ Occurrence
  - ▪ Timing
  - ▪ Effects
  - ▪ Location
- ❏ Goals
  - ▪ Offensive
  - ▪ Defensive
  - ▪ Nonintervention

		4th edition
Defining Criteria	page 6	148–171
Identification	page 7	55, 100, 237

Tactical Objectives

- ☐ Rescue
    - Risk assessment
    - Event profile
    - Public protective actions
    - Isolation and control
    - Evacuation
    - Protect in place
- ☐ Spill control
    - Absorption
    - Adsorption
    - Covering
    - Damming
- ☐ Diking
    - Dilution
    - Diversion
    - Dispersion
    - Retention
    - Vapor dispersion
    - Vapor suppression
- ☐ Leak control
    - Neutralization
    - Overpacking
    - Patching and plugging
    - Pressure isolation and reduction
    - Solidification
    - Vacuuming
- ☐ Fire control
    - Foam application
    - Other extinguishing application
    - Exposure protection
- ☐ Recovery
    - Product removal and transfer
    - Termination procedures

		4th edition
Identification	page 7	100
Tasks to Be Performed	page 78	321
PPE Required for the Mission	page 15	55, 100
Classification	page 7	148–171
Verification	page 7	205
Resource Coordination	page 19	308
Termination	page 30	433

Threat Hazard Relationships

- What is the profile of the incident and how does it influence the tactical objectives and strategic goals?
    - Develop an incident action plan that addresses multijurisdictional /multi-agency goals
    - Consolidated command structure to achieve the desired objective
    - Ensure the tactical plan lends itself to achieving the identified strategic goals
- Ensure the clean sweep of the operational area for secondary/tertiary devices
- Do the tactical conditions match the strategic goals within the context of the operation?
    - Rescue operations
    - Public protective actions
    - Spill control (confinement)
    - Leak control (containment)
    - Fire control
    - Recovery

THE EIGHT STEP PROCESS©/6. Implement Response Objectives

7. DECONTAMINATION

Decontamination Site Selected

- ❏ Site selection
  - Topography and wind direction
  - Area sloped toward entrance
  - Ground contours considered
- ❏ Entry and exits well marked
- ❏ Location is identified as the warm zone
- ❏ Water source established
  - Run-off considered
  - Decontamination solutions identified
    - Decon solutions for people
    - Decon for equipment
- ❏ Disposal containers placed
  - Equipment
  - Personal belongings
  - Evidence
- ❏ Area of refuge identified
- ❏ Mass decontamination different than responders
  - Capacity identified
- ❏ Decontamination sectorization
  - Responder decontamination
  - Public/civilian decontamination
  - Supply officer identified

Resources Established

- ❏ PPE identified from information branch
- ❏ Decontamination equipment secured
  - Supportive equipment
    - Water source
    - Containment equipment
    - Ancillary equipment placed
  - Human resource
    - Positions covered for each level
    - Medical personnel available (see Evaluation of Decontamination Personnel)

Size-up Evaluation

- Emergency decontamination procedures briefed to all members prior to entry
- Establish areas of safe refuge for emergency conditions
- Consider entry options
  - Rescue and reconnaissance
  - Mitigation techniques
  - Tactical engagement (SWAT)
  - Crime scene investigation
- Mass decontamination
  - Capacity to handle (resource intensive)
  - Cold weather decontamination
  - Gender considerations
  - Ethnic considerations
- Effectiveness of the decontamination process
- Medical care/surveillance upon exit

Threat Hazard

Target Analysis
Operational Conditions
- Number of victims versus resources
- Chemical properties
- Human exposures
- Limit decontamination numbers by public protective actions

Risk Assessment
- Law enforcement containment
- Temperature (present/future)

Probability Assessment
- Effect on social structure
- Behavior of responders
- Behavior of victims

■ Decontamination

- ❏ Chemical effects of the chemical(s) evaluated
  - Information branch has identified
    - Reactivity to water
    - Reactivity to decon solutions
- ❏ Personnel showering facilities identified
  - On site
    - Emergency decontamination area identified
  - Off site
- ❏ Decontamination solution containment
  - Containment procedures
  - Permitted into sewer system
- ❏ Equipment in position
  - Decon team in position
  - Decon procedures briefed
    - Emergency decon
    - Area of refuge
  - Stations identified
    - Clean side, dirty side
    - Entrance point
    - Technical decon
    - SCBA removal
    - Removal and isolation of PPE
    - Removal of personal clothing
    - Body wash
- ❏ Dry off and don clean clothing available
  - Medical evaluation
    - Assign to a responsible functional group

■ Evaluation of Decontamination Personnel

- ❏ Health effect of exposure
  - Signs and symptoms of chemical exposure
  - Medical section for responders
    - EMS preparation for response personnel
    - Emergency decon and responders assessment
  - Medical section for public
    - EMS preparation for civilian patients (victims)
  - Hospital notification
    - Personnel decontaminated
    - Victims decontaminated
    - Number of potential victims
  - Medical surveillance
    - Decontamination Steps 8, 9 evaluation
    - Injured
      Minimal treatment in contaminated area
      Maintain Basic Life Support
      Remove and isolate contaminated clothing
      Advanced Life Support when available

Interrelated Strategic and Tactical Objectives

Smoke: © Greg Henry/ShutterStock, Inc.; Flames: © Photos.com

7. DECONTAMINATION

GOAL: To ensure the safety of both emergency responders and the public by reducing the level of contamination on-scene and minimizing the potential for secondary contamination beyond the incident scene.

FUNCTION: Decontamination (decon) is the process of making personnel, equipment, and supplies "safe" by reducing or eliminating harmful substances (i.e., contaminants) that are present when entering and working in contaminated areas (i.e., the hot zone or inner perimeter). Although decon is commonly addressed in terms of "cleaning" personnel and equipment after entry operations, response personnel should remember that in some instances, due to the nature of the materials involved, decontamination of clothing and equipment may not be possible and these items may require disposal.

All personnel trained to the First Responder Operations level should be capable of delivering an emergency decon capability. At most "working" hazmat incidents, decon services will be provided by HMRTs or fire and rescue units working under the direction of a Hazmat Technician. Questions pertaining to disposal methods and procedures should be directed to environmental officials and technical specialists based on applicable federal, state, and local regulations.

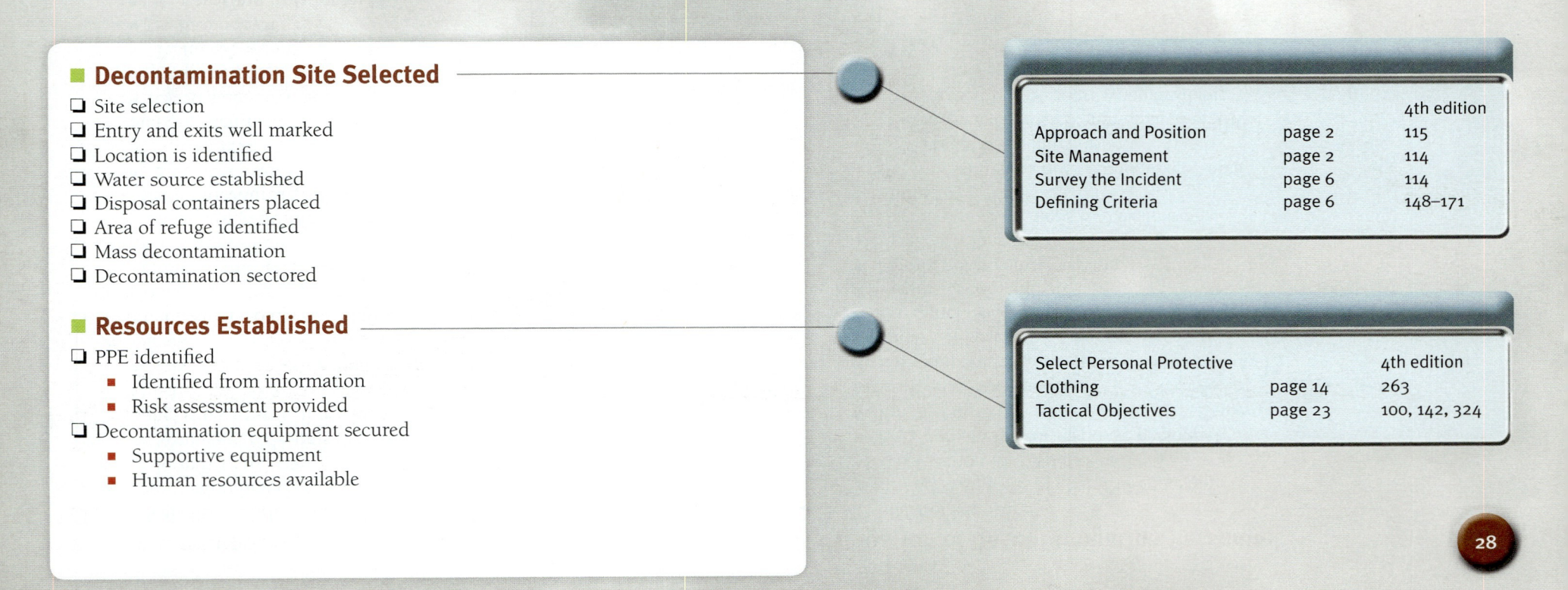

■ **Decontamination Site Selected**

❏ Site selection
❏ Entry and exits well marked
❏ Location is identified
❏ Water source established
❏ Disposal containers placed
❏ Area of refuge identified
❏ Mass decontamination
❏ Decontamination sectored

		4th edition
Approach and Position	page 2	115
Site Management	page 2	114
Survey the Incident	page 6	114
Defining Criteria	page 6	148–171

■ **Resources Established**

❏ PPE identified
  ▪ Identified from information
  ▪ Risk assessment provided
❏ Decontamination equipment secured
  ▪ Supportive equipment
  ▪ Human resources available

Select Personal Protective		4th edition
Clothing	page 14	263
Tactical Objectives	page 23	100, 142, 324

Decontamination Site Setup
- ❏ Chemical effects evaluated
- ❏ Personnel showing facilities
- ❏ Decontamination solution containment
- ❏ Equipment in position
- ❏ Dry off and don clean clothing

Evaluation of Decontamination
- ❏ Health effects of exposure
  - Signs and symptoms of the chemical identified
  - Medical section for responders
  - Medical section for public
  - Hospital notification
  - Medical surveillance
- ❏ PIO contacted for media

		4th edition
Resource Coordination	page 19	308
Decontamination	page 26	395

		4th edition
Exposure Assessment	page 10	33, 142, 243
Termination of the incident	page 30	433

Threat Hazard Relationships
- Ensure strong perimeter controls
- Strong management of the decontamination operations
  - Access control points
  - Operational security
  - Established personnel with appropriate credentials
  - Separate levels of decontamination
    - Public decontamination (victims)
    - Response team decontamination
    - Hazardous materials team
    - SWAT team
    - Crime scene investigation
      - Evidence recovery and processing
- Forced protection dependent on the operation
- Consistent policing of the decontamination corridor(s) for secondary or suspicious devices
- Integrated information dissemination from within the incident intelligence plan
- Evaluation and reevaluation of the hazards presented (pages 10, 11)

THE EIGHT STEP PROCESS©/7. Decontamination

8. TERMINATING THE INCIDENT: RESTORATION AND RECOVERY

Scene Termination

❏ Active analysis—hazard/risk assessment
- Monitoring and detection results
- Reconnaissance analysis
- Confinement analysis

❏ Evaluation of incident stabilization by Safety and Hazmat Group leader
- Needs assessment of each branch
- Level of service that each functional area can provide
- Resources
  - Notification and evaluation of internal resources
  - Notification and evaluation of external resources
- Identify equipment/supplies
  - Disposal
  - Decontamination
  - Re-supply

Responsibility Transfer

❏ Identification/notification of external resources

❏ Brief owners and contractors of incident
- Initial nature of the incident
- Actions taken during the incident
  - Concerns during the incident
  - Considerations of the incident
  - Chemicals involved
  - Hazard/risk assessment
    - Initial assessment
    - Current assessment
- Identify the agency or authority assuming control of incident

Incident Debriefing (<15 minutes)

❏ Identify to the responders the hazards that were found

❏ Identify the signs and symptoms of exposure to the chemical(s)

❏ Identify damaged or expended equipment/supplies

❏ Identify conditions of the site at the time of transfer

❏ Assign information-gathering responsibilities
- Postincident analysis
- Critique

❏ Identify the need for critical incident stress management

Considerations

- Maintain control of the perimeter
- Control long periods of inactivity during the termination phase
- Look at the environmental impact and provide appropriate notifications
- Address health concerns of the public and responders
- Assess stress placed on all surrounding containers before releasing them to contractors
- Evaluate the low areas and surrounding confined spaces for flammable, reactive, toxic environments
- Ensure the site safety plan is followed during termination
- Ensure contractors are able to manage the level of incident you are handing them
- Give a media debriefing, describing the incident, agencies involved, and personnel utilized

Threat Hazard

Target Analysis

Operational Conditions
- Equipment review
- Security of scene maintained
- Briefing of relieving agency
- Immediate problems requiring attention
- Assess future requirements

Risk Assessment
- Ensure health information has been identified and distributed
- Intentional release—identify the area isolated

Probability Assessment
- Consider external threat conditions
  - Intentional release
  - Appropriate authorities notified
  - Review situational status
- Contractors cleared by authorities

Postincident Analysis

- ❏ Documentation of incident
  - Notification of appropriate regulatory agency
  - Determine financial responsibility
  - Identify stress/behavior event
  - Probable cause—accident investigation
- ❏ Command and control
  - Incident Management System established
  - IMS in accordance with the existing ERP/SOP
  - Chain of command followed
  - Strategic objectives translated into effectively managed operational task assignments
  - Tactical objectives understood by the responder(s) who implemented them
- ❏ Tactical operations
  - Operations occurred in a timely manner
  - Obstacles that impeded operational times or implementation
    - Specific tasks
    - Identified functions
- ❏ Resources
  - Level or resources available
  - Used effectively
    - Mutual aid
    - Automatic response
  - Support services
    - Level of provisions
- ❏ Plans and procedures
  - Conflicting issues within the ERP/SOP
  - Phases of the event covered by the ERP/SOP
  - Roles and responsibilities appropriately identified
- ❏ Training issues
  - Reinforcement of current training initiatives
  - Advanced training required
  - Interagency training initiatives
  - Additional training required

Critique

- ❏ Formalized structure (time limited) identifying:
  - Incident presentation and mitigation strategies
  - Hazard/risk assessment
  - Incident management strategies
    - Specific tasks
    - Identified functions
  - Incident constraints and obstacles
    - Resources available
    - Functions dependent on resources
    - Educational component/impact
    - Size of incident
    - Level of incident (containment systems)
  - Recommendations from each branch
    - Information
    - Reconnaissance/entry
    - Resources
    - Hazmat medical
    - Decontamination
  - Participant
    - Roles and activities
    - Operational constraints
  - Operations
    - Structured review of operations
    - Identifying branch/sector operations
    - Operational constraints
  - Group
    - Operational concerns
    - Avenues for operational enhancement
    - External constraints

THE EIGHT STEP PROCESS©/8. Terminating the Incident: Restoration and Recovery

Interrelated Strategic and Tactical Objectives

Smoke: © Greg Henry/ShutterStock, Inc.; Flames: © Photos.com

8. TERMINATING THE INCIDENT: RESTORATION AND RECOVERY

GOAL: To ensure that overall command is transferred to the proper agency when the emergency is terminated, and that all postincident administrative activities are completed per local policies and procedures.

FUNCTION: This is the termination of emergency response activities and the initiation of post-emergency response operations (PERO), including investigation, restoration, and recovery activities. This would include the transfer of command to the agency that will be responsible for coordinating all post-emergency activities.

■ Scene Termination

❏ Active analysis
- Monitoring and detection
- Reconnaissance
- Containment
- Confinement
- Hazard behavior

❏ Evaluation of incident stabilization
- Needs assessment of each branch
- Level of service that can be provided
- Identify resources required for service

■ Responsibility Transfer

❏ Identification/notification of external resource
- Notify all resources of termination stage

❏ Brief owners/contractors about incident
- Actions taken during the incident
- Concerns about the incident
- Considerations of the incident
- Chemicals involved
- Hazard risk assessment

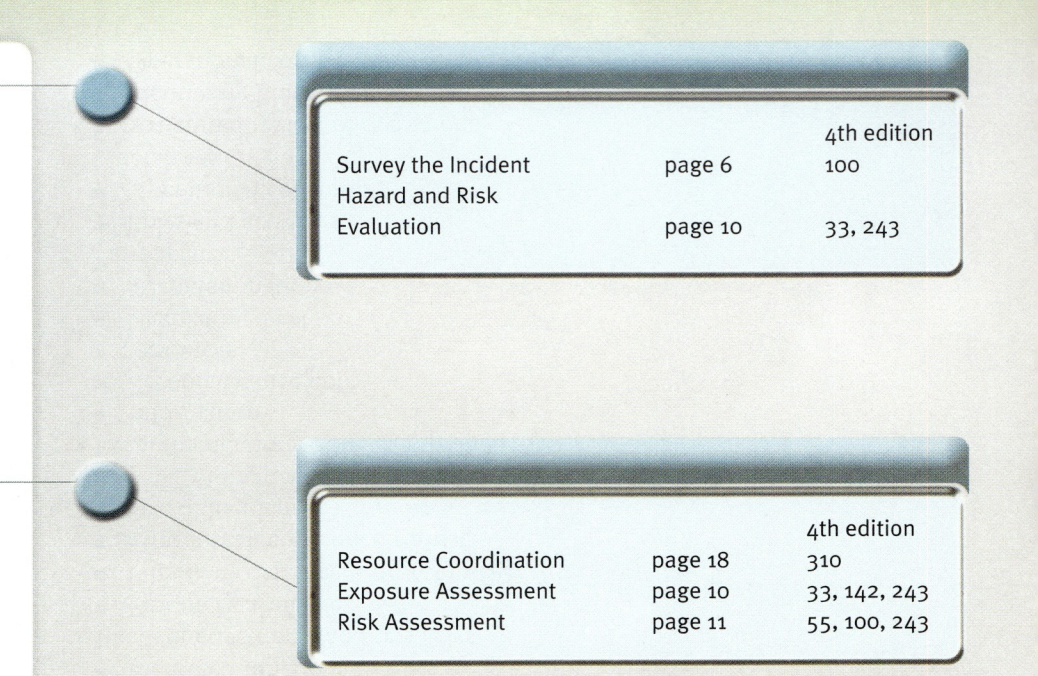

		4th edition
Survey the Incident	page 6	100
Hazard and Risk Evaluation	page 10	33, 243

		4th edition
Resource Coordination	page 18	310
Exposure Assessment	page 10	33, 142, 243
Risk Assessment	page 11	55, 100, 243

Incident Debriefing

- ☐ Identify the hazards found
- ☐ Identify the signs and symptoms of exposure
- ☐ Identify damaged or expended equipment/supplies
- ☐ Identify the conditions of the site
- ☐ Assign information-gathering responsibilities
- ☐ Identify need for critical incident stress management

Postincident Analysis

- ☐ Documentation of the incident
- ☐ Command and control
- ☐ Tactical operations
- ☐ Resources
- ☐ Plans and procedures
- ☐ Training issues

Critique

- ☐ Formalized structure
  - Incident presentation
  - Hazard/risk assessment
  - Constraints and obstacles
  - Recommendations
- ☐ Incident management strategies
- ☐ Participant
- ☐ Overall operations
- ☐ Group operations

Implementing Response		4th edition
Objectives	page 22	324
Exposure Assessment	page 10	32, 142, 243
Risk Assessment	page 11	55, 100, 243

Review of Steps 1–8

Review of Steps 1–7

Threat Hazard Relationships

- Establish recovery security action plan
- Maintain perimeter control based on:
  - Intelligence
  - Evidence recovery
  - Continued operational objectives
- Ensure restoration and recovery objectives
  - Multijurisdictional/multiagency recovery plan
  - Resource needs

THE EIGHT STEP PROCESS©/8. Terminating the Incident: Restoration and Recovery

First Response Actions

SECTION 2

- First Responder Initial Action
- Placards and Labels
- Transportation Overview: Damage Assessment
- Cargo Tank Trucks/Containers
- Railroad Tank Cars
- Nonbulk Packaging
- Storage Tanks
- Markings and Colors
- CBRNE Matrix

INTERRELATED SCENE ACTION PLAN

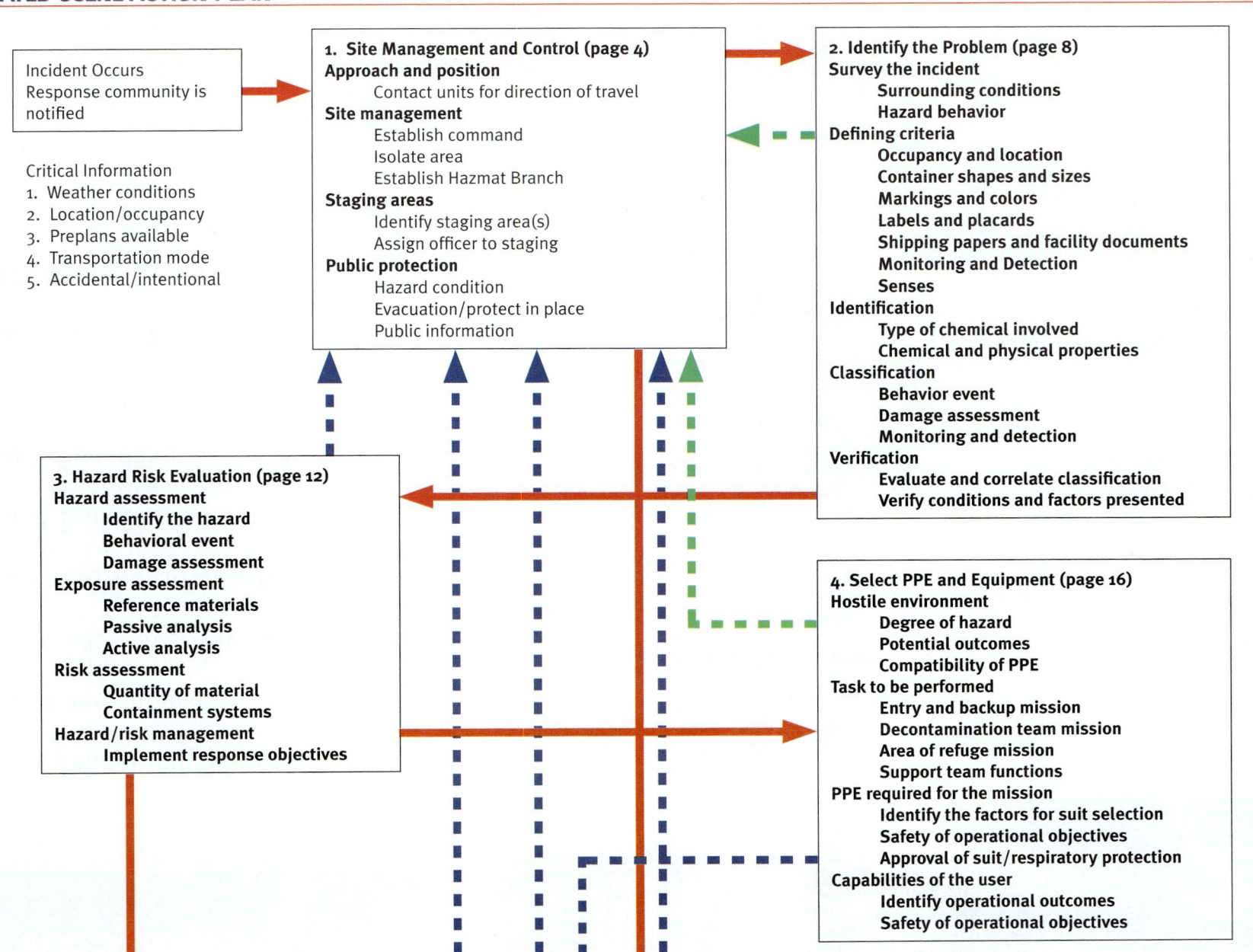

Incident Occurs
Response community is notified

Critical Information
1. Weather conditions
2. Location/occupancy
3. Preplans available
4. Transportation mode
5. Accidental/intentional

1. Site Management and Control (page 4)
Approach and position
        Contact units for direction of travel
Site management
        Establish command
        Isolate area
        Establish Hazmat Branch
Staging areas
        Identify staging area(s)
        Assign officer to staging
Public protection
        Hazard condition
        Evacuation/protect in place
        Public information

2. Identify the Problem (page 8)
Survey the incident
        Surrounding conditions
        Hazard behavior
Defining criteria
        Occupancy and location
        Container shapes and sizes
        Markings and colors
        Labels and placards
        Shipping papers and facility documents
        Monitoring and Detection
        Senses
Identification
        Type of chemical involved
        Chemical and physical properties
Classification
        Behavior event
        Damage assessment
        Monitoring and detection
Verification
        Evaluate and correlate classification
        Verify conditions and factors presented

3. Hazard Risk Evaluation (page 12)
Hazard assessment
        Identify the hazard
        Behavioral event
        Damage assessment
Exposure assessment
        Reference materials
        Passive analysis
        Active analysis
Risk assessment
        Quantity of material
        Containment systems
Hazard/risk management
        Implement response objectives

4. Select PPE and Equipment (page 16)
Hostile environment
        Degree of hazard
        Potential outcomes
        Compatibility of PPE
Task to be performed
        Entry and backup mission
        Decontamination team mission
        Area of refuge mission
        Support team functions
PPE required for the mission
        Identify the factors for suit selection
        Safety of operational objectives
        Approval of suit/respiratory protection
Capabilities of the user
        Identify operational outcomes
        Safety of operational objectives

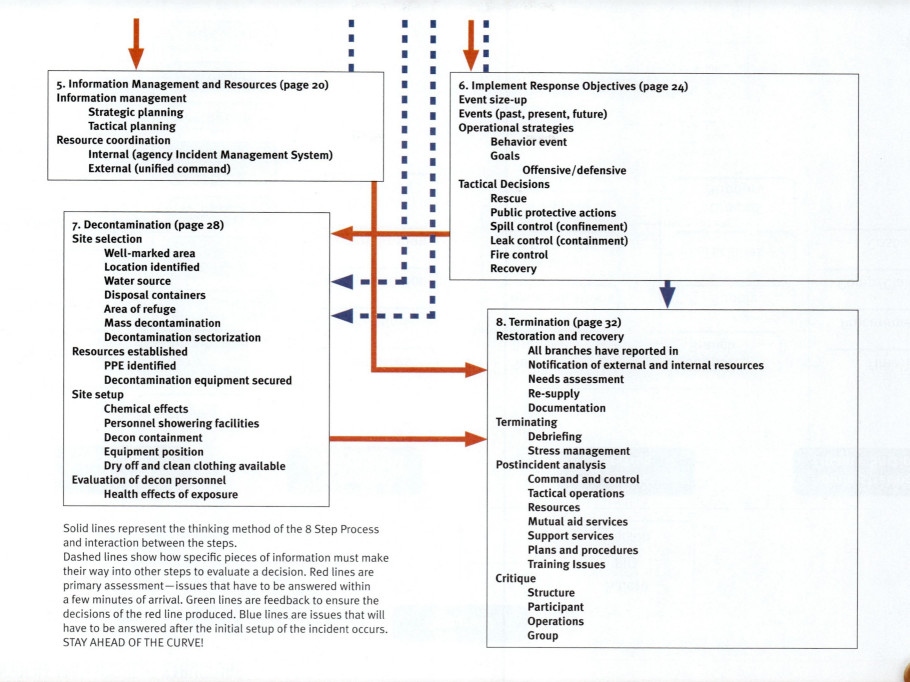

FIRST RESPONSE ACTIONS/*Interrelated Scene Action Plan*

INCIDENT MANAGEMENT STRUCTURE

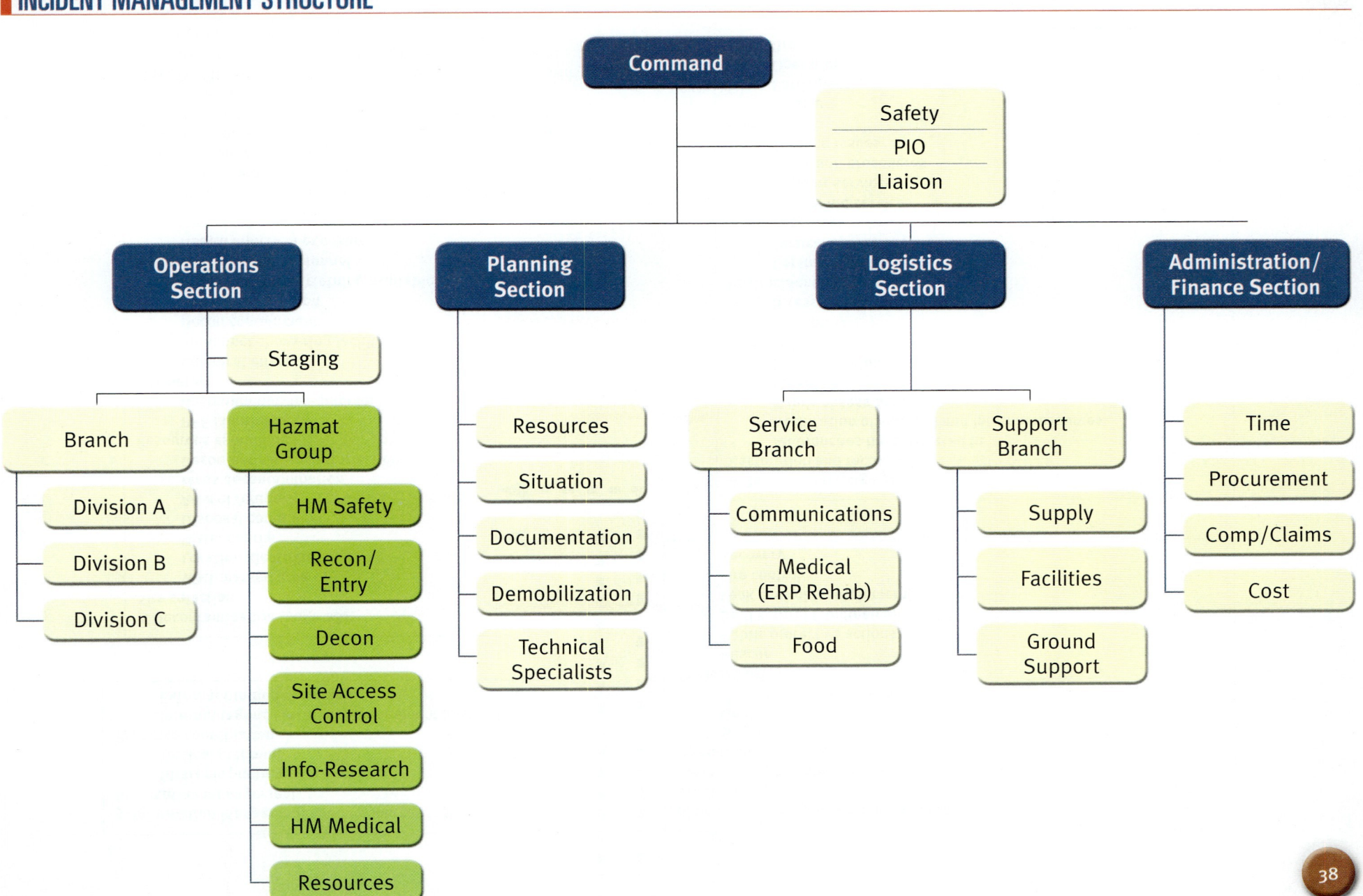

HAZMAT GROUP STAFFING

```
                    Hazmat Group
                    Supervisor
                         |
                      Safety
                         |
    ┌────────────┬───────┼────────┬────────────┐
Information  Recon/Entry Resources  Hazmat    Decon
 Research                           Medical
```

Information Research

Tasks:
Data gathering
Coordination

Evaluate:
Hazards/risk
Public protection
Develop action plans

Recon/Entry

Tasks:
Entry/backup
Recon
Monitoring
Sampling

Evaluate:
Hazards/risk
Develop action plans

Resources

Tasks:
Control and tracking of supplies and equipment

Evaluate:
With identified hazards and projected risks, coordinates with logistics

Hazmat Medical

Tasks:
Pre-post entry physicals
Technical medical guidance

Evaluate:
With identified hazards and projected risks, identify the medical management plan
Develop medical plan

Decon

Tasks:
Research and development of decontamination plan
Set up and evaluation of effective decon

Evaluate:
Hazards/risk
Need for additional decontamination resources for:
    ERT
    Civilians
    Equipment

FIRST RESPONSE ACTIONS / *Hazmat Group Staffing*

INCIDENT ACTION PLAN: GOALS

1. Site Management and Control

❑ Command structure established
- Command identified
- Approach to incident relayed to units

❑ Isolation perimeters
- Hazard control (deny entry)
  - 1,000 ft for flammable/toxic release
  - 300 ft for suspicious device

❑ Evacuation/protect in place
- Immediate area
- High occupancy
- Special occupancy

2. Identify the Problem

❑ Evaluate surrounding conditions
- Clouds/plume
- Fire/smoke

❑ Occupancy/location

❑ Container shapes and sizes

❑ Markings and colors

❑ Shipping papers/facility documents

❑ Behavior event
- Stress to the container
- Breach of the container
- Release
- Engulfment
- Impingement
- Harm factors
  - Thermal
  - Radiation
  - Asphyxiation
  - Toxic
  - Corrosive
  - Etiological
  - Mechanical

❑ Damage assessment
- Crack
- Gouge/score
- Dent/burn

First Responder Initial Actions

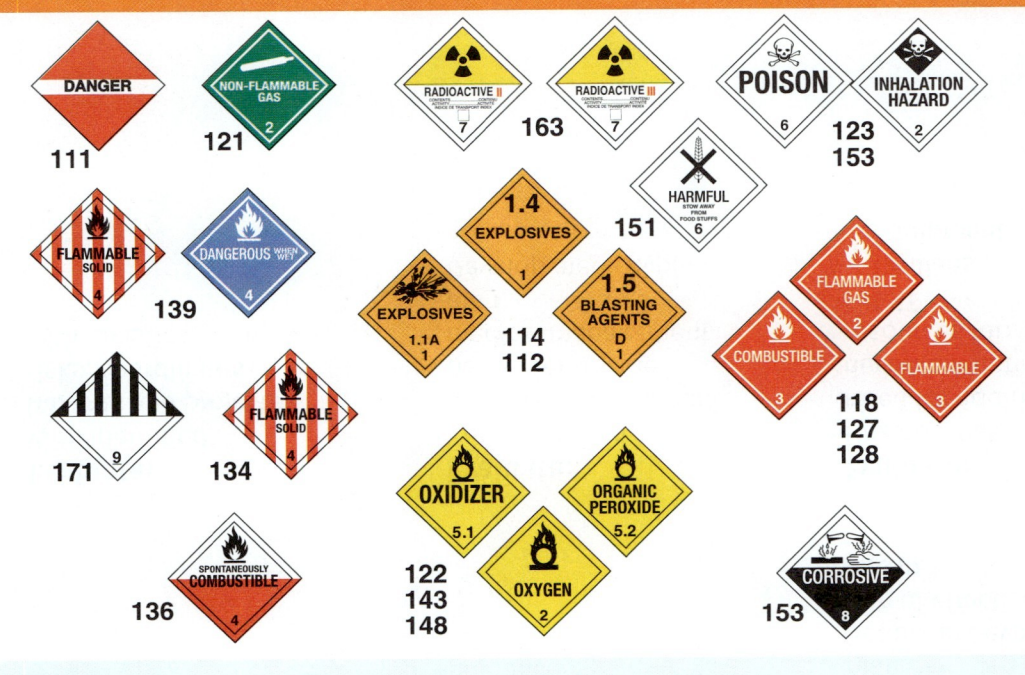

General Guide DOT 111	Isolate spill or leak 330–660 feet in all directions If tank, railcar, or truck involved in fire, Isolate 1/2 mile and evacuate Firefighter gear offers limited protection in fire situations only
Explosive Guide DOT 112	Isolate spill or leak 1/3 mile in all directions If tank, railcar, or truck involved in fire, Isolate 1 mile and evacuate Firefighter gear offers limited protection in fire situations only
Etiological Guide DOT 158	Isolate spill or leak 30–80 feet in all directions Keep all unauthorized personnel away Firefighter gear offers limited protection in fire situations only
Radiological Guide DOT 164	Isolate spill or leak 80–160 feet in all directions If tank, railcar, or truck involved in fire, Isolate 1000 feet and evacuate Firefighter gear offers limited protection in fire situations only

3. Hazard and Risk Evaluation
- [ ] Identification of the hazard(s)
  - Risk assessment
  - Quantity of material: _____
  - State of matter: solid, liquid, gas
  - Vapor pressure

4. Select PPE
- [ ] Entry is based upon present gear and level of protection afforded
- [ ] Identification of hazard must be made

5. Information Management
- [ ] Relay information assessed to command
- [ ] Relay which objective cannot be achieved

6. Implement Response Objectives
- [ ] Defensive
  - Spill control/confinement
    - Absorption
    - Covering/damming/diking
    - Dilution
    - Diversion/dispersion
    - Retention
    - Vapor suppression/dispersion
  - Leak control/containment
    - Neutralization
  - Fire control
    - Extinguishing agent
    - Water supply
    - Reactive chemicals
  - Specialized operations—support functions
    - Clan lab
    - EOD support
    - Confined space
    - SWAT support

7. Decontamination
- [ ] Site selection—location identified
- [ ] Simple decon—engine company with hose line
- [ ] Mass decontamination
  - Water supply
  - Uphill, run-off away from responders
  - Hose lines and master streams
  - Area of refuge identified
  - Evaluation of decontamination

8. Termination
- [ ] Evaluation of incident
- [ ] Level of stabilization

Incident Location:_____ Date:_____ Time:_____

Type of Incident:_____ Placard Class:_____

UN ID:_____ Green Section:_____ DOT Guide #:_____ NIOSH Page #:_____

State of Matter : S L G Water Reactive: Y N Oxidizer Y N Radioactive Y N

State Found : S L G Time:_____ Temp:_____ Humidity:_____ Dew Point:_____

Manufacture:_____ BP:_____ IDLH:_____ VP:_____

IT:_____ UEL:____

Contact Person:_____ VD:_____ LEL:____

TLV (PEL):_____

Contact#:_____

FIRST RESPONSE ACTIONS / *First Responder Initial Actions*

Placards

Hazard Class	Placard		Placard Name	Application

1

Explosives
1.1 Explosives with mass explosion hazard
1.2 Explosives with a projection hazard
1.3 Explosives with predominantly a fire hazard
1.4 Explosives with no significant blast hazard
1.5 Very insensitive explosives; blasting agents
1.6 Extremely insensitive detonating agents

The designated alphabetical letter is used to categorize different types of explosive substances and materials for the purpose of stowage and segregation. Canada classifies certain gases as a corrosive gas (anhydrous ammonia).

2

Gases
2.1 Flammable gas
2.2 Nonflammable compressed gas
2.3 Gases toxic by inhalation
2.4 Corrosive gas (Canada)

3

**Flammable Liquids
(and Combustible Liquids — United States)**

4

Flammable Solids: Spontaneously Combustible Materials, and Dangerous When Wet
4.1 Flammable solids
4.2 Spontaneously combustible materials
4.3 Dangerous when wet materials

5  **Oxidizers and Organic Peroxides**
5.1 Oxidizers
5.2 Organic peroxides

6 **Toxic Materials and Infectious Substances**
6.1 Toxic materials
6.2 Infectious materials

7  **Radioactive Materials**

Labels will carry additional information (contents—name of radionuclide, radioactive activity, transport index). All labels will show vertical bars indicating radioactive levels (I, II, III) overprinted in red on lower half of each label.

8 **Corrosive Materials**

9 **Miscellaneous Dangerous Goods**
9.1 Miscellaneous dangerous goods (Canada)
9.2 Environmentally hazardous substances (Canada)
9.3 Dangerous wastes (Canada)

Material that presents a hazard during transport, but does not meet the definition of any other hazard class.

Transportation Overview: Damage Assessment

Pressure Tank Car

Non-pressure Cargo Tank
MC 306 / DOT 406

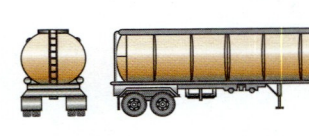

Corrosive Cargo Tank
MC 312 / DOT 412

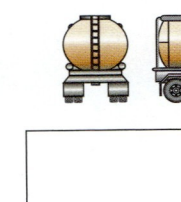

Low-Pressure Cargo Tank
MC 307 / DOT 407

Corrosive Tank Car

Pressure Cargo Tank
MC 331

Cryogenic Tank Car

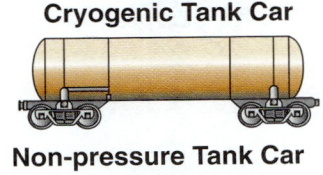

Cryogenic Liquid Tank
MC 338

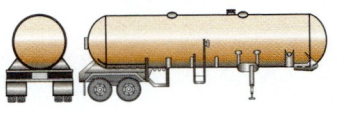

Non-pressure Tank Car

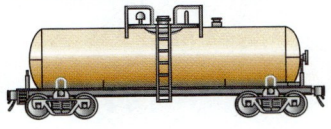
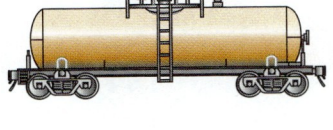

Compressed-Gas Trailer

Back — Right Side — Front

Back — Front

Front — Left Side — Back

Container on a Flat Car

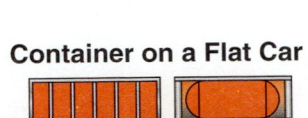

Intermodals

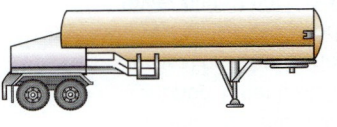

Trailer on a Flat Car (TOFC)

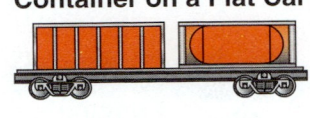

Box Car

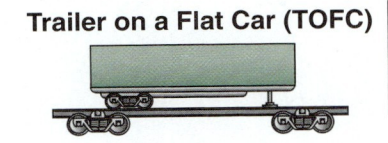

Gondola

Covered Hopper Car

High-Pressure Specialized Car

44

Corrosive Liquid Cargo Tank MC 312/DOT 412

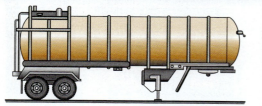

- Circular cross section, smaller diameter with external reinforcing ribs often visible
- May be found in double-shell configuration
- Insulated tanks may not appear circular in cross section
- Overturn and splash protection at dome cover/valve locations
- 5,000–6,000 gallons capacity

Common Commodities

Corrosive liquids and high-density liquids

Pressure Cargo Tank MC 331

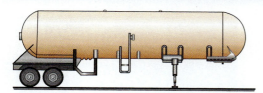

- Circular cross section with rounded ends or heads
- Design pressures of not less than 100 psi or more than 500 psi
- Single-shell, noninsulated tank
- Upper two thirds painted white or highly reflective color
- Capacity ranges from 2,500 gallons (bob tail) to 11,500 gallons (cargo)

Common Commodities

LPG
Anhydrous ammonia
Propane

FIRST RESPONSE ACTIONS/*Transportation Overview: Damage Assessment*

Cryogenic Liquid Tank MC 338

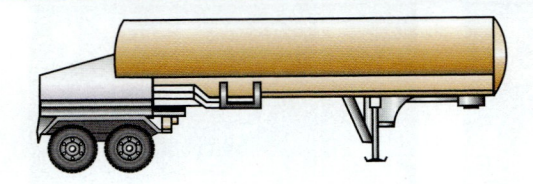

Courtesy of Jack B. Kelly, Inc.

- Well-insulated steel tank, thermos bottle design with flat tank ends
- Operating pressure 22 psi
- Typically marked as refrigerated liquid
- Double-shell tank with vapor discharge from relief valve
- Normal relief from valves

Common Commodities

Liquefied oxygen
Liquefied nitrogen
Liquefied carbon dioxide

High-Pressure Tube Trailers

Courtesy of Jack B. Kelly, Inc.

- Multiple staked cylinders with over-pressure regulator for each cylinder
- Operating pressure 3,000–5,000 psi
- Protected valves assembly in rear
- Cylinders are stacked and manifolded in rear

Common Commodities

Corrosive liquids and high-density liquids

Non-pressure Cargo Tank MC 306 / DOT 406

- Elliptical cross section; flat end oval cross section indicates nonpressurized tank (less than 3 psi)
- Rollover protection
- Bottom valves
- Multiple compartments
- Flammable or combustible cargo
- 9,000 gallons capacity

Common Commodities

Gasoline
Fuel oils
Alcohol

Low-Pressure Cargo Tank MC 307 / DOT 407

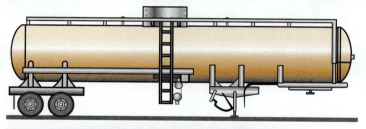

- With or without skin; without skin, stiffing rings visible
- Rubber-lined or double-steel construction
- Circular cross section with low pressures (up to 40 psi)
- Rollover protection
- Flammable or combustible cargo
- 6,000–7,000 gallons capacity

Common Commodities

Flammable liquids
Combustible liquids
Mild acids and bases
Poisons

Dry Bulk Commodity Carriers

Courtesy of Polar Tank Trailer L.L.C.

- Carry very heavy loads
- Capacity up to 1,500 cubic ft
- Use air pressure to transfer loads
- Static charges are common hazards

Common Commodities

Ammonium nitrate
Cement
Dry caustic soda
Plastic pellets

Ton Containers

- Test pressures 500–1,000 psig
- Transported by truck or rail
- Container valves are found at one end under a protective cap
- Transport one ton of chlorine: actual cylinder weighs approximately 1,800 lb empty
- Transport chlorine, sulfur dioxide, phosgene, and refrigerant gases
- Chlorine and sulfur dioxide containers have fusible plugs on each head

Non-pressure Tank Containers — IM-101 / IM-102

- Maximum allowable working pressure for IM101 (IMO Type 1) ranges from 25.4 to 100 psig
- Maximum allowable working pressure for IM102 (IMO Type 2) ranges from 14.5 to 25.4 psig
- Transported by all modes
- Capacities up to 6,300 gallons
- Often use a combination pressure/vacuum relief device; rupture disks may also be found
- Transport hazardous materials and food-grade commodities

Pressure Tank Containers — DOT Spec 51 / Spec 51L

- Working pressures up to 100 psig
- Have a circular cross section; as large as 6 ft in diameter and 20 ft long
- Capacities range up to 5,500 gallons
- Transported by all modes
- Transport of LPG, anhydrous ammonia, bromine, sodium and aluminum alkyls

Specialized Tank Containers — Cryogenic IMO Type 7

- Approved use by DOT special exemption
- Tank-within-a-tank design with insulation between inner and outer tanks
- Cryogenic liquids

Specialized Tank Containers — Tube Modules

- Multiple staked cylinders with over-pressure regulator for each cylinder
- Operating pressure 3,000–5,000 psi
- Cylinders are stacked and manifolded

FIRST RESPONSE ACTIONS / *Transportation Overview: Damage Assessment*

"Super Sacks" and Totes: Intermediate Bulk Containers (IBCs)

Bulk bags or "super sacks": Preformed packaging, constructed of flexible materials (e.g., polypropylene), available plain or coated, or with liners.
Standard sizes range from 15 to 85 cubic ft with capacities of 500 to 5,000 lb.
Transports solid materials such as fertilizers, pesticides, and water treatment chemicals.

Portable bins: Approximately 4 ft by 6 ft high. May contain up to 7,700 lb. Loaded through the top and unloaded from the side or bottom.
Transports solid materials, such as ammonium nitrate fertilizer, calcium carbide, and other hazardous materials.

Non-pressure portable tank (totes): May have rectangular or circular cross sections approximately 6 ft high.
Capacity of approximately 300 to 400 gallons. May contain internal pressures up to 100 psi.
Used for transport of liquid materials such as liquid fertilizer, water treatment chemicals, and flammable solvents.

Radioactive Protective Overpacks and Casks

Also referred to as Type A and Type B packaging.
Used for the transportation of radioactive materials and radioactive waste.

Type A: Designed to prevent loss or dispersal of contents under normal conditions of transport.

Type B: Meets same criteria as Type A but designed to meet standards for performance under hypothetical accident conditions. Consists of rigid metal materials with a cylindrical or box-like configuration.

Pressure Tank Car

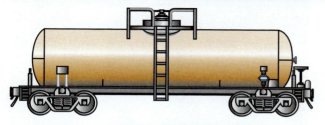

- Cylindrical tank with rounded ends
- Fittings and valves enclosed in dome
- Off-white paint indicates sprayed-on thermal insulation
- Black paint usually indicates a jacketed tank car
- Tank pressures range from 100 to 600 psi
- Capacities of 4,000–45,000 gallons

Common Commodities

Flammable, nonflammable, and poisonous gases

Pressure tank car classes
DOT 105   DOT 114
DOT 109   DOT 120
DOT 112

Cryogenic Liquid Tank Car

- Well-insulated "thermos bottle" design
- Double-shell tank similar to fixed storage tanks
- Transport low-pressure refrigerated liquids (pressures 25 psig or lower)
- Absence of any top fittings
- Loading/unloading and safety relief device often found in cabinets at diagonal corners or on one end at ground level
- Cryogenic tank car classes DOT 113 and AAR 204W

Common Commodities

Cryogenic liquid oxygen, liquid hydrogen, liquid nitrogen, liquid argon

Non-pressure Tank Car

Courtesy of private source

- Horizontal tank with flat or nearly flat ends
- Fittings and valving visible on top of car
- Older cars will have an expansion dome with visible fittings
- Tank pressures range from 35 to 100 psi
- Capacities of 4,000–45,000 gallons

Non-pressure tank classes

DOT 103   AAR 201
DOT 104   AAR 203
DOT 111   AAR 206
DOT 115   AAR 211

Some non-pressurized cars are now being constructed with housing around all fittings.

Common Commodities

Flammable liquids, poisons, oxidizers, molten solids, and some liquefied gases

Corrosive Liquid Tank Car

- Similar to non-pressure car
- Multiple fittings
- Tank pressures range from 35 to 100 psi
- Capacities of 4,000–45,000 gallons
- Often have bottom unloading values

Common Commodities

Corrosives

52

Covered Hopper Car

- Usually covered with bottom unloading
- Transports fertilizer, grain, and plastic pellets
- Sometimes used for oxidizers

Container on Flat Car (COFC)

- Traditional highway cargo vans on special flat car
- Highway cargo tanks not permitted
- Transports mixed loads of all hazard classes

FIRST RESPONSE ACTIONS/*Transportation Overview: Damage Assessment*

Gondola Car

© Christina Richards/Shutterstock

- Usually uncovered with low sides
- Transports bulk ores and other solid materials
- Sometimes used for LSA

Box Car

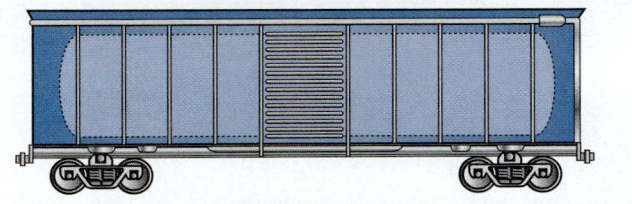

- Enclosed steel or wood inside
- Used for general freight
- Transports drums, cylinders, boxes, and other nonbulk containers

© Arthur Eugene Preston/Shutterstock

Trailer on Flat Car (TOFC)

- Traditional highway cargo vans on special flat car
- Highway cargo tanks not permitted
- Transports mixed loads of all hazard classes

Nonbulk Packaging

Drums

- Cylindrical packaging made of metal, plastic, fiberboard, plywood, or other suitable materials.
- Typical capacity is 55 gallons.
- Overpack drums used to hold damaged or leaking nonbulk packaging have 85-gallon capacity.
- Have removable (open head) or non-removable heads (tight or closed head).
- Closed-head drum usually contains two openings—2 inch and ¾ inch diameter plugs or "bungs."
- Used for liquids and solids—for example, lubricating grease, caustic powders, corrosive powders, flammable solvents, spontaneously combustible materials, and poisons.

Cylinders

- Typically made of steel, although aluminum or fiberglass aluminum may be found.
- Found in a variety of sizes (e.g., 20 lb propane cylinder to 1 ton chlorine cylinder).
- Do not have a uniform taper on cylinder head; thread design will vary depending upon contents.
- Used for pressurized and liquefied gases, such as acetylene, LPG, chlorine, and oxygen.

Insulated Cylinders

- Consist of an insulated metal cylinder contained within an outer protective metal jacket.
- Area between cylinder and jacket is normally under vacuum.
- Designed for a specific range of service pressures and temperatures.
- Found in a ranges of sizes.
- Used for cryogenic liquids, such as liquefied argon, helium, nitrogen, and oxygen.

Bags

- Flexible packaging constructed of cloth burlap, Kraft paper, plastic, or a combination of these materials.
- Closed by folding and gluing, heat sealing and stitching, crimping with metal, or twisting and tying.

Bottles

- Constructed of glass and plastic, although metal and ceramic are sometimes used.
- Closed with threaded caps or stoppers.
- Range from ounces to 20 gallons.
- Usually placed in an outside packaging for transport.
- Sometimes referred to jars or jugs.
- Used for liquids and solids, including laboratory reagents, corrosive liquids, and light-sensitive and-reactive materials.

Boxes

- Rigid packaging that completely encloses the contents.
- Commonly used as the outside packaging for other nonbulk packages. Inner packaging may be surrounded with absorbent or vermiculite.
- Constructed of fiberboard, wood, metal, plywood, plastic, or other suitable materials.
- Fiberboard boxes may contain boxes up to 550 lb.
- Used for solid and liquids; all categories of hazardous materials can be found in this packaging.
- Commonly used for radioactive, etiological agents, and infectious samples.

Multicell Packaging

- Consists of a form-fitting, expanded polystyrene box encasing one or more bottles.
- When transporting certain hazardous materials, maximum bottle capacity is 4 liters; up to 6 bottles may be placed in one multiple package.
- Used for specialty chemicals, corrosive liquids, and various solvents.

Carboys

- Glass or plastic "bottles" that may be encased in outer packaging (e.g., polystyrene box, wooden crate, plywood drum).
- Range in capacity to more than 20 gallons.
- Used for liquids such as acids, caustics, and water.

FIRST RESPONSE ACTIONS/*Nonbulk Packaging*

Atmospheric and Low-Pressure Liquid Storage Tanks

Cone Roof Tank

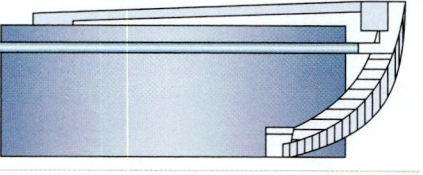

- Flammable liquids
- Combustible liquids
- Corrosives

Open Floating Roof

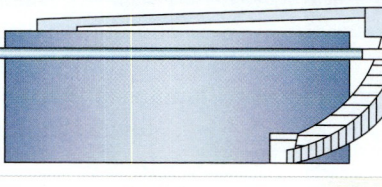
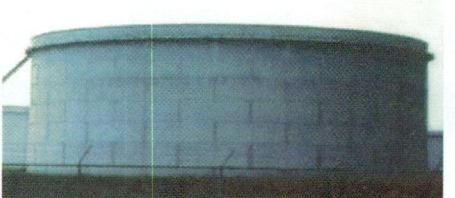

- Flammable liquids
- Combustible liquids

Covered Floating Roof

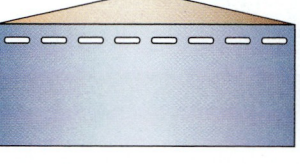

- Flammable liquids
- Combustible liquids

Horizontal Tank

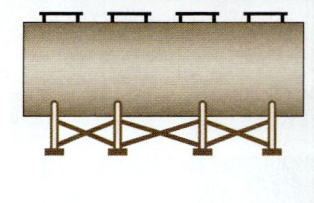

- Low-pressure tank
- Flammable liquids
- Combustible liquids
- Corrosives
- Fertilizers
- Chemical solvents

Vertical Tank

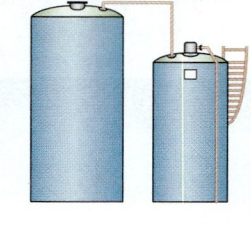

- Flammable liquids
- Combustible liquids
- Corrosives
- Fertilizers
- Chemical solvents

High-Pressure Horizontal Tank

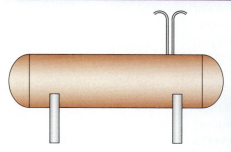

- Process chemicals
- Anhydrous ammonia
- LPG

High-Pressure Spherical Tank

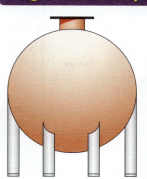

- Process chemicals
- Anhydrous ammonia
- LPG

Cryogenic Liquid Storage Tank

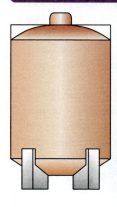

- Liquid oxygen
- Liquid nitrogen
- Liquid carbon dioxide
- Liquid argon
- Liquid hydrogen

FIRST RESPONSE ACTIONS / *Pressurized Storage Vessels*

Railroad Tank Car Specification Marking System

DOT   111   A   60   AL   W   1

Other Car Features:
Fittings, Materials, Linings

Type of Weld Used:
"W" Fusion Welding (most common)
"F" Forge Welding

Type of Material Used in Tank Construction:

"No Letter"	Carbon Steel
"AL"	Aluminum (Classes 103, 106, 109, 111)
"A-AL"	Aluminum Alloy
"N"	Nickel
"C", "D", or "E"	Stainless Steel (alloy steel)

Tank Test Pressure (PSI)

Seperator Character:
Significant only for Class 105, 112, 113, 114 Tank Cars and some 111 Tank Cars when refitted.

"A"	Top and bottom shell couplers
"S"	Tank head shield, top and bottom shelf couplers
"J"	Jacketed thermal protection, tank head shields, and top and bottom shelf couplers
"T"	Spray-on thermal protection, tank head shields, top and bottom shelf couplers

Authorizing Agency:
Tank car specifications start with three letters designating the agency under whose authority the specifications was issued:

DOT—Department of Transportation

AAR—Association of American Railroads

ICC—Interstate Commerce Commission (assumed by DOT in 1966)

CTC—Canadian Transport Commission

TC—Transport Canada (replacing CTC)

Class Designation:
The three-digit class designation follows the authorizing agency:

NONPRESSURE TANK CARS		CRYOGENIC LIQUID TANK CARS	
DOT 103	AAR 201	DOT 113	AAR 204W
DOT 104	AAR 203		AAR 204XT (inside boxcar)
DOT 111	AAR 206		
DOT 115	AAR 211	MISCELLANEOUS TANK CARS	
PRESSURE TANK CARS		DOT 106 Multiunit Tank Car Tanks (Ton Containers)	
DOT 105	DOT 114	DOT 110 Multiunit Tank Car Tanks (Ton Containers)	
DOT 109	DOT 120	DOT 107 High-Pressure Tank Car	
DOT 112		AAR 207 Pneumatically Unloaded Covered Hopper	

Railroad Tank Car Markings

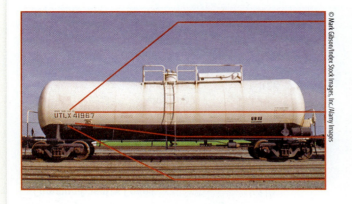

Reporting Marks and Car Number
"X" Indicates a railcar is not owned by the railroad (lack of an "X" indicates railroad ownership)
"Z" Indicates a trailer
"U" Indicates a container

Load Limit in Pounds and Kilograms

Empty Weight of Car in Pounds and Kilograms

Month and Year of Lightweight Date

DOT Car Specification Marking

Water Capacity of Tank (on pressure cars) in pounds and kilograms

Tank Test and Safety Valve Test Information

FIRST RESPONSE ACTIONS / *Railroad Tank Car Markings*

Intermodal Portable Tank Markings

UPTU   660355   6

US 2275

IMO-1   R.I.D./A.D.R.

IM 101-C.T.C. IMPACT APPROVED

SP3250

CSC   AAR 600   FRA

TIR

43.5 P.S.I (3 BAR)   M.A.W.P.

Specification Marking; IM 101, IM 102, Spec. 51

DOT Exemption Marking—authorized DOT exemption marked DOT-E followed by an exemption number (DOT E8623)

AAR-600 Marking for interchange purpose in rail transportation—indicates tanks that can be used for regulated materials

Reporting marks and number—International Container bureau registered in France

Country, Size, and Type Markings

Common Country Codes

BM (BER)	Bermuda	LIB	Liberia
CH (CHS)	Switzerland	NLX	Netherlands
DE	Germany	NZX	New Zealand
DKX	Denmark	PA (PNM)	Panama
FR (FXX)	France	PIX	Philippines
GB	Great Britain	PRC	People's Republic of China States
HKXX	Hong Kong	RCX	People's Republic of China (Taiwan)
ILX	Israel		
IXX	Italy	SGP	Singapore
JP (JXX)	Japan	SXX	Sweden
KR	Korea	US (USA)	United States

Common Size Codes

20 = 20 ft (8 ft high)
22 = 20 ft (8 ft 6 in high)
24 = 20 ft (>8 ft 6 in high)

Common Type Codes
Nonhazardous M.A.W.P.
Bar Test Pressure
70 = <0.44 (6.4 psig)
71 = 0.44 − 1.47 (21.3 psig)
72 = 1.47 − 2.94 (42.6 psig)
73 = Spare

Common Type Codes
Hazardous M.A.W.P.
Bar Test Pressure
74 = <1.47 (21.3 psig)
75 = 1.47 − 2.58 (37.4 psig)
76 = 2.58 − 2.94 (42.6 psig)
77 = 2.94 − 3.93 (57.0 psig)
78 = >3.93
79 = Spare

International ADR (Road) Kemler Code

First Digit
1. Explosive
2. Gas
3. Flammable liquid
4. Flammable solid
5. Oxidizer
6. Toxic
7. Radioactive
8. Corrosive
9. Miscellaneous danger

X = Do not use water

Second and Third Digits
0. No additional hazard
1. Explosive risk
2. Gas produced in contact with water
3. Flammable risk
4. Molten state (elevated temperatures)
5. Oxidizing risk
6. Toxic risk
8. Corrosive risk
9. Risk of violent reaction from spontaneous combustion or self-polymerization

ADR Code

X432

2059

UN Number

Select Specific Special Numbers

20	Inert gas	339	Flammable liquid with violent reaction	80	Corrosive
22	Refrigerated gas	39	Flammable liquid with violent reaction	X80	Corrosive, reacts with water
223	Refrigerated flammable gas	40	Flammable solid	88	Highly corrosive
225	Refrigerated oxidizing gas	X423	Flammable liquid, reacts with water	89	Corrosive, violent reaction
23	Flammable gas	44	Flammable solid, molten state	90	Miscellaneous danger
236	Toxic flammable gas	446	Toxic flammable solid, molten state		
239	Flammable gas (violent reaction)	46	Toxic flammable solid		
25	Oxidizing gas	50	Oxidizer		
26	Toxic gas	539	Organic peroxide		
265	Oxidizing toxic gas	558	Corrosive oxidizer		
266	Highly toxic gas	559	Oxidizer, violent reaction		
268	Corrosive toxic gas	589	Corrosive oxidizer, violent reaction		
286	Toxic corrosive gas	60	Toxic		
30	Flammable liquid	70	Radioactive material		
X323	Flammable liquid, reacts with water	72	Radioactive gas		
33	Highly flammable liquid	723	Flammable radioactive gas		
X333	Flammable liquid, reacts with water	73	Radioactive liquid		
336	Toxic flammable liquid	74	Radioactive flammable solid		
337	Corrosive flammable liquid	75	Radioactive oxidizer		
X338	Corrosive flammable, reacts with water	76	Radioactive toxic		

Military Marking System

Fire Division Symbols

Class 1 * Division 1
Mass Detonation

Class 1 *
Division 12 Explosion
with Fragment Hazard

Class 1 * Division 3
Mass Fire Hazard

Class 1 * Division 4
Moderate Fire Hazard

Chemical Hazard Symbols

Highly Toxic Chemical
Agents Set no. 1

Harassing Agents
Set no. 2

White Phosphorus
Munitions
Set no. 2

Apply No Water

Wear Protective Mask
(or Breathing Apparatus)

Shipping Papers Information

Mode of Transportation	Title of Shipping Papers	Location of Shipping Papers	Responsible Persons
Highway	Bill of lading	Cab of vehicle	Driver
Rail*	Waybill/consist	With crew (conductor)	Crew (conductor)
Water	Dangerous cargo manifest	Wheelhouse or pipeline captain or master	Container on barge
Air	Airbill with shippers certification for restricted article	Cockpit	Pilot

*Standard Transportation Commodity Code (STCC) number is used extensively on rail transportation shipping papers.

Proper Shipping Name—Identifies the name of the hazmat as found in the DOT Hazardous Materials Regulations. The word "WASTE" will precede the proper shipping name for those shipments that are classified as hazardous wastes.

DOT Hazard Class/Division Number—Indicates the material's primary and secondary (as appropriate) hazards as listed in the DOT Hazardous Materials Regulations. A division is a subset of a hazard class. NOTE: A hazmat may meet the criteria for more than one hazard class, but is assigned to only one hazard class.

Identification Number(s)—The four-digit identification number assigned to each hazardous material. The identification number may be found with the prefix "UN" (United Nations) or "NA" (North America).

Packing Group—Further classifies hazardous materials based on the degree represented by the material. There are three groups:

- Packing Group I—Indicates great danger
- Packing Group II—Indicates medium danger
- Packing Group III—Indicates minor danger

Packing groups may be shown as "PGI" and are assigned to Class 2 (compressed gases), Class 7 (Radioactives), and sometimes Division 6.2 (infectious substances and ORM-D materials).

FIRST RESPONSE ACTIONS/*Shipping Papers and Facility Documents*

Total Quantity—Indicates the quantity by net or gross mass, capacity, or other unit. May also indicate the type of packaging. The number and type of packaging may be entered on the beginning line of the shipping description. Carriers often use abbreviations to indicate the type of packaging. Union Pacific Railroad examples:

BA = Bale	CA = Case	CH = Covered Hopper	DRM = Drum	KIT = Kit	SAK = Sack
BG = Bag	CAN = Can	CL = Carload	JAR = Jar	KL = Container Load	TB = Tube
BOX = Box	CR = Crate	CY = Cubic Yard	JUG = Jug	PA = Pail	TC = Tank Car
BC = Bucket	CTN = Carton	CYL = Cylinder	KEG = Keg	PKG = Package	TL = Trailer Load

Emergency Contact—Indicates the telephone number for the shipper or shipper's representative that may be accessed 24/7 in the event of an accident. CHEMTREC may be displayed as the emergency contact.

Shipping Papers: Additional Entries

- Compartment Notation—Identifies the specific compartment of a multicompartmented rail car or cargo tank truck in which the material is located. On railcars, compartments are numbered sequentially from the "B" end (the end where the hand brake wheel is located), while cargo tank trucks are numbered sequentially from the front.

- Residue (Empty Packaging)—Identifies packaging that contains a hazmat residue and has not been cleaned and purged or reloaded with a material that is not subject to the DOT hazardous materials regulations. Residue is indicated by the words "Residue: Last Contained" before the proper shipping name. It is only used in rail transportation.

- HOT—Identifies elevated-temperature materials other than molten sulfur and molten aluminum.

- Technical Name—Identifies the recognized chemical name currently used in scientific and technical handbooks, journals, and texts. General descriptors may be found provided that they identify the general chemical group.

- Not Otherwise Specified (N.O.S.) Notations—If the proper shipping name of a material is an "N.O.S" notation, the technical name of the hazardous materials must be entered in parentheses with the basic description. If the material is a mixture or solution of two or more hazardous materials, the technical names of at least two components that predominantly contribute to the hazard of the mixture/solution must be entered.

- Subsidiary Hazard Class—Indicates a hazard of a material other than the primary hazard assigned.

- Reportable Quantity (RQ) Notation—Indicates the material is a hazardous substance as designated by the EPA. The letters "RQ" (reportable quantity) must be shown either before or after the basic shipping description entries. This designation indicates that any leakage of the substance above its RQ value must be reported.

- Marine Pollutant—Indicates that the material meets the definition of a marine pollutant.

- EPA Waste Stream Number—Indicates the number assigned to a hazardous waste stream by the USEPA to identify that waste stream.

- EPA Waste Characteristic Number—Indicates the general hazard characteristics assigned to a hazardous waste by EPA; these characteristics include EPA Corrosivity, EPA Toxicity, EPA Ignitability, and EPA Reactivity.

- **Radioactive Material Information**—Indicates the following:
    - "Radioactive Material"— if not part of the proper shipping name
    - Name of each radionuclicide
    - Physical/chemical form
    - Activity in curies
    - Transport Index (if applicable)
    - U.S. Department of Energy Approval Number (if applicable)
    - Fissile Exempt (if applicable)
    - Fissile Class (if applicable)
    - Label applied
- **Poison Notation**—Indicates that a liquid or solid material is poisonous when the fact is not disclosed in the shipping name.
- **Poison-Inhalation Hazard (PIH) Notation**—Indicates gases and liquids that are poisonous by inhalation.
- **Hazard Zone**—Indicates relative degree of hazard in terms of toxicity (only appears for gases and liquids that are poisonous by inhalation).
    - Zone A—LC_{50} less than or equal to 200 ppm (most toxic)
    - Zone B—LC_{50} greater than 200 ppm but less than or equal to 1,000 ppm
    - Zone C—LC_{50} greater than 1,000 ppm but less than or equal to 3,000 ppm
    - Zone D—LC_{50} greater than 3,000 ppm but less than or equal to 5,000 ppm (least toxic)
- **Dangerous When Wet Notation**—Indicates a material that, when in contact with water, is liable to become spontaneously flammable or give off flammable or toxic gas at a rate greater than 1 liter per kilogram of the material per hour.
- **Limited Quantity (LTD QTY)**—Indicates a material being transported in a quantity for which there is a specific labeling and packaging exception.
- **Canadian Information**—Indicates information required for hazardous materials entering or exiting Canada in addition to that required in the United States.
- **Placard Notation**—Indicates the placard applied to the container. Where placards are not required, the notation "MARKED" is followed by the four-digit identification number.
- **Placard Endorsement**—Indicates the presence of a hazardous material requiring a placard. Found inside a rectangle made with any symbol (*, $, #).
- **Trade Name**—The name that enables organizations to access the MSDS for additional information.
- **Hazardous Materials STCC Number**—A seven-digit Standard Transportation Commodity Code (STCC) number will be found on all shipping papers accompanying rail shipments of hazmats. It is also found when intermodal containers are changed from rail to highway movement. The first two digits ("49") are the key identifier. The STCC will follow the notation "HAZMAT STCC."
- **Shipper Contact**—Indicates the identity of the producer or consolidator of the materials described.

The API Color Code Chart

GASOLINE

DISTILLATES

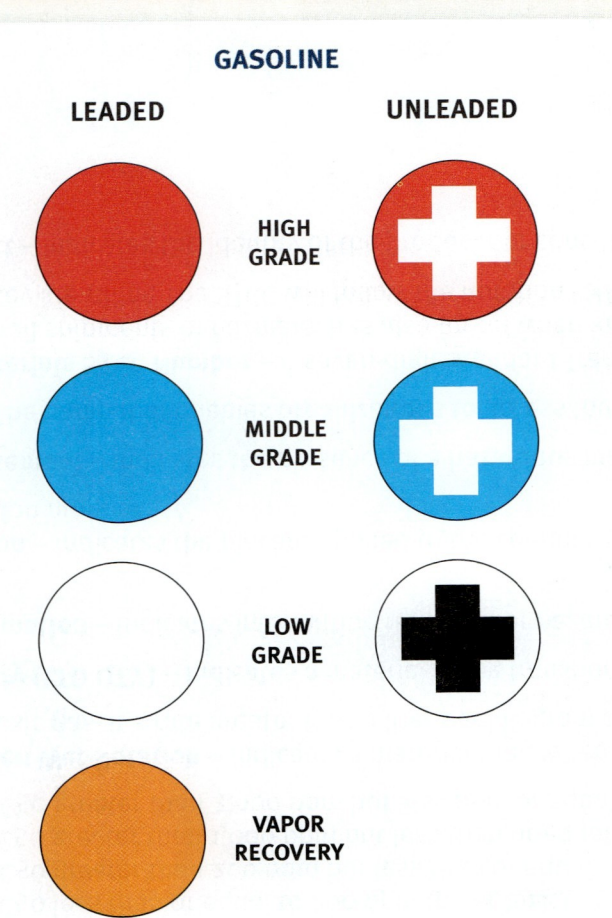

EXAMPLES OF SYMBOLS FOR PRODUCTS WITH EXTENDERS (OPTIONAL)

UNLEADED HIGH-GRADE GASOLINE WITH EXTENDER

DIESEL WITH EXTENDER

UN Marking of Packaging

| 1 | 2 | 3 | 4 | 5 | 6 | 7 | 8 | 9 |

1. United Nations symbol or letters "UN"

2. Packaging identification codes for nonbulk packaging consist of the following:
   - Number identifying the type of packaging:
     - "1" means a drum
     - "2" means a wooden barrel
     - "3" means a jarrican
     - "4" means a box
     - "5" means a bag
     - "6" means composite packaging
     - "7" means a pressure receptacle

   - Capital letter indicating the material of construction:
     - "A" means steel (all types and surface treatments)
     - "B" means aluminum
     - "C" means natural wood
     - "D" means plywood
     - "F" means reconstituted wood
     - "G" means fiberboard
     - "H" means plastic
     - "L" means textile
     - "M" means paper, multiwall
     - "N" means metal (other than steel or aluminum)
     - "P" means glass, porcelain, or stoneware

   - A second numeral indicating the category of packaging within the type to which the packaging belongs:
     - "1" indicates a nonremovable head (drum)
     - "2" indicates a removable head (drum)

   - For a composite packaging, two capital letters are used in sequence in the second position of the code:
     - First letter indicates the material of the inner receptacle
     - Second letter indicates the material of the outer packaging

DRUMS
1A1 Nonremovable-head steel drum
1A2 Removable-head steel drum
1B1 Nonremovable-head aluminum drum
1B2 Removable-head aluminum drum
1N1 Nonremovable-head metal drum
    (other than steel or aluminum)
1N2 Removable-head metal drum
    (other than steel or aluminum)
1D Plywood drum
1G Fiber drum
1H1 Nonremovable-head plastic drum
1H2 Removable-head plastic drum

BARRELS
2C1 Bung-type wooded barrel
2C2 Slack-type (removable head)
    wooden barrel

JERRICANS
3A1 Nonremovable-head head steel jerrican
3A2 Removable-head steel jerrican
3H1 Nonremovable-head plastic jerrican
3H2 Removable-head plastic jerrican

BOXES
4A1 Unlined and uncoated steel box
4A2 Steel box with inner liner or coating
4B1 Unlined and uncoated aluminum box
4B2 Aluminum box with inner liner or coating
4C1 Ordinary natural wooden box
4C2 Natural wood box with sift-proof walls
4D Plywood box
4F Reconstituted wood box
4G Fiberboard box
4H1 Expanded plastic box
4H2 Solid plastic box

BAGS
5H1 Unlinded and uncoated woven plastic bag
5H2 Sift-proof woven plastic bag
5H3 Water-resistant woven plastic bag
5H4 Plastic film bag
5L1 Unlined or uncoated textile bag
5L2 Sift-proof textile bag
5L3 Water-resistant textile bag
5M1 Multiwall paper bag
5M2 Multiwall water-resistant paper bag

COMPOSITE PACKAGINGS
6HA1 Plastic receptacle within a protective steel drum
6HA2 Plastic receptacle within a protective steel crate or box
6HB1 Plastic receptacle within a protective aluminum drum
6HB2 Plastic receptacle within a protective aluminum crate or
    box
6HC1 Plastic receptacle within a protective wooden box
6HD1 Plastic receptacle within a protective plywood drum
6HD2 Plastic receptacle within a protective plywood box
6HG1 Plastic receptacle within a protective fiber drum
6HG2 Plastic receptacle within a protective fiberboard box
6HH Plastic receptacle within a protective plastic drum
6PA1 Glass, porcelain, or stoneware receptacle within a
    protective steel drum
6PA2 Glass, porcelain, or stoneware receptacle within a
    protective steel crate or box
6PB1 Glass, porcelain, or stoneware receptacle within a
    protective aluminum drum
6PB2 Glass, porcelain, or stoneware receptacle within a
    protective wooden box
6PC Glass, porcelain, or stoneware receptacle within a
    protective wooden box

---

3. Performance Standard
"X" for packaging meeting Packing Group I, II, and III tests
"Y" for packaging meeting Packing Group II and III tests
"Z" for packaging meeting Packing Group III tests

---

4. Specific gravity/mass designation (in kilograms)

---

5. Hydrostatic test pressure, in kilopascals (kPa) to the nearest 10 kPa
"S" indicates use for solids or inner packaging

---

6. Year of manufacture (last two digits)

---

7. State authorizing the mark (e.g., USA)

---

8. Name and address or symbol of the manufacturing agency or approval agency

---

9. Minimum thickness, in millimeters (mm), for metal or plastic drums, or jerricans intended for reuse or reconditioning

NFPA 704 Marking System

The NFPA 704 Marking system distinctively indicates the properties and potential dangers of hazardous materials. The following is an explanation of the meanings of the Quadrant Numerical Codes.

Flammability — (Red)

Susceptibility to burning is the basis for assigning degrees within this category. The method of attacking the fire is influenced by this susceptibility factor.

- 4 Very flammable gases or very volatile flammable liquids. Shut off flow and keep cooling water stream on exposed tanks or containers.
- 3 Materials that can be ignited under almost all normal temperature conditions. Water may be ineffective because of low flash point.
- 2 Materials that must be moderately heated before ignition will occur. Water spray must be used to extinguish the fire because the material can be cooled below its flash point.
- 1 Materials that must be preheated before ignition can occur. Water may cause frothing if it gets below the surface of the liquid and turns to steam.
- 0 Materials that will not burn.

Health — (Blue)

In general, health hazard in firefighting involves a single exposure that may vary from a few seconds to an hour.

- 4 Materials too dangerous to health to expose fire fighters. A few whiffs of the vapor cloud could cause death or the vapor of liquid could be fatal on penetrating the fire fighter's normal full protective clothing.
- 3 Materials extremely hazardous to health, but areas may be entered with extreme care. No skin surface should be exposed.
- 2 Materials hazardous to health, but areas may be entered freely with full-face mask and self-contained breathing apparatus, which provides eye protection.
- 1 Materials only slightly hazardous to health. It may be desirable to wear self-contained breathing apparatus.
- 0 Materials that would offer no hazard beyond that of ordinary combustible materials upon exposure under fire conditions.

Special information (white)
Materials that demonstrate unusual activity

Reactivity (Stability) (Yellow)

The assignment of degrees in the reactivity category is based upon the susceptibility of materials to release energy either by themselves or in combination with water; fire exposure was one of the factors considered along with conditions of shock and pressure.

- 4 Materials that (in themselves) are readily capable of detonation or of explosive decomposition or explosive reaction at normal temperatures and pressure; includes materials that are sensitive to mechanical or localized thermal shock. If a chemical with this hazard rating is in an advanced or massive fire, the area should be evacuated.
- 4 Materials that (in themselves) are capable of detonation or of explosive decomposition or explosive reaction but that require a strong initiating source or that must be heated under confinement before initiation.
- 4 Materials that (in themselves) are normally unstable and rapidly undergo violent chemical change but do not detonate. Includes materials that can undergo chemical change with rapid release of energy at normal temperatures and pressures or that can undergo violent chemical change at elevated temperatures and pressures. Also includes those materials that may react violently with water or that may form potentially explosive mixtures with water. In advanced or massive fires, firefighting should be done from a safe distance or from a protected location.
- 4 Materials that (in themselves) are normally stable but that may become unstable at elevated temperatures and pressures or that may react with water with some release of energy but will not react violently. Caution must be used in approaching the fire and applying water.
- 0 Materials that (in themselves) are normally stable even under fire exposure conditions and that are not reactive with water. Normal firefighting procedures may be used.

CBRNE Signs and Symptoms Matrix: Chemical Matrix

	1	2	3	4	5	6	7	8	9	10	11	12	13	14	15	16
Scene Assessment																
Fruity smell						O	O									
Geraniums odor					O											
Irritating burning	O			O											O	O
Faint, fishy, or musty			O										O			O
Garlic/horseradish odor		O														
Oil-looking droplets		O	O		O		O									
Mist in the air	O	O	O		O	O	O	O	O		O				O	O
Bitter almond odor												O				
Gas cloud								O	O		O	O				
Freshly mown hay									O							
Immediate fatalities							O					O	O	O		
Pool smell										O						
Patient Assessment Respiratory																
Labored breathing	O	O	O	O	O	O	O	O	O	O	O	O	O	O	O	O
Stridor	O			O												
Wheezing								O	O	O	O					O
Tachypnea												O	O	O		
Cardiovascular																
Chest pain	O			O		O	O	O	O	O	O	O	O	O		O
Tachycardia	O			O	O	O	O			O		O	O	O		
Bradycardia						O	O									
Awareness Status																
Restlessness	O		O	O	O	O	O			O	O			O	O	
Seizures/convulsions						O	O					O	O	O		
Coma						O	O								O	

1	Corrosive	Acids/Bases
2		Sulfur Mustard
3		Nitrogen Mustard
4		Phosgene Oxime
5		Lewisite
6	Neurotoxin	Organophosphates
7		Nerve Agent
8	Irritants	Choking Agents
9		Phosgene
10		Chlorine
11		Anhydrous Ammonia
12		Cyanide
13		Hydrogen Sulfide
14		Carbon Monoxide
15		Nitrates/Nitrites
16	Riot Control Agents	Mace/Pepper Spray

Symptom	1	2	3	4	5	6	7	8	9	10	11	12	13	14	15	16
Eyes																
Tearing, severe				O		O	O			O	O					O
Pinpoint pupils						O	O									
Edema	O			O						O	O				O	
Irritation	O	O	O	O	O		O	O	O	O	O	O	O	O		O
Grit feeling in eye		O	O	O												
Soapy feeling on skin	O															
Eye burning	O	O	O	O	O					O	O					
Eye tearing	O	O	O	O	O			O	O	O	O	O	O			O
Photophobia	O	O	O	O	O	O	O	O	O	O	O					O
Eye pain	O			O		O	O			O	O					O
Ear, Nose, Throat																
Excessive secretions		O	O		O	O	O									O
Moderate secretions		O	O	O	O	O		O	O	O	O					
Coughing						O	O					O	O		O	O
Skin																
Excessive sweating						O	O									
Reddening of skin	O	O	O	O	O					O	O					O
Immediate needle pain					O											
Intense pain	O			O	O					O	O					O
Blisters (short term)	O		O		O					O	O					
General																
Fasciculations						O	O									
Nausea/vomiting		O	O	O	O	O	O			O	O	O	O	O		O
Defecation						O	O									
Muscle weakness						O	O							O	O	
Muscle twitching												O	O	O		
Cramps of skeletal muscles						O	O									
Anxiety				O	O	O	O	O	O	O	O	O	O	O	O	
Weakness						O	O							O	O	

FIRST RESPONSE ACTIONS / CBRNE Signs and Symptoms Matrix: Chemical Matrix

CBRNE Signs and Symptoms Matrix: Biological Matrix

Legend:
1 Anthrax · 2 Brucellosis · 3 Cholera · 4 Glanders · 5 Plague · 6 Tularemia · 7 Typhus · 8 Q Fever · 9 Smallpox · 10 VHF · 11 VEE · 12 Yellow Fever · 13 Botulinum · 14 Ricin · 15 T2 · 16 SEB

	1	2	3	4	5	6	7	8	9	10	11	12	13	14	15	16
Incident Assessment																
Increase in flu-type symptoms	O	O	O		O	O	O	O	O	O	O	O		O		O
Respiratory infections/complaints	O	O			O	O	O	O	O							O
Rapid increase in fatal infectious cases					O				O	O						
Similar disease manifestations from one location	O	O	O		O	O	O	O	O	O	O	O	O	O	O	O
Patients are presenting with:																
Flu/chills	O	O	O		O	O	O	O	O		O					O
Weakness/uncoordination		O	O				O				O		O	O	O	
Infections	O	O			O		O		O	O						
Fever	O	O			O	O	O	O	O	O	O	O		O		O
Confusion/dizziness	O								O	O	O		O		O	
Shortness of breath	O			O	O		O					O	O	O	O	
Bloody sputum				O	O				O					O		
Patient Assessment Respiratory																
Cough	O	O			O	O		O	O		O			O	O	O
Pneumonia			O	O	O			O		O						
Pulmonary edema				O					O					O	O	O
Upper airway cough (stridor)	O	O			O											
Cyanosis	O				O								O	O		
Gastrointestinal																
Abdominal pain	O	O	O		O				O	O				O	O	
Nausea/vomiting		O	O		O			O	O	O	O	O	O	O	O	O
Diarrhea	O	O	O		O					O	O			O		O
Blood within the vomit										O				O	O	
Bloody feces/tarry stools	O									O		O		O	O	

	1	2	3	4	5	6	7	8	9	10	11	12	13	14	15	16
Skin																
Bruising					O					O						
Raised skin lesions									O							
Rash		O				O	O		O	O					O	
Blisters									O				O		O	
Ulcers	O			O		O									O	
Bruising (bursting of the capillary beds—petechiae)					O						O					
Red skin									O							
Cardiovascular																
Chest pain	O	O				O		O							O	O
Awareness Status (General)																
Joint pain/muscle pain	O	O		O					O	O	O	O	O	O		
Uncoordination											O				O	
Headache		O			O	O			O	O	O	O	O			O
Nosebleeds										O					O	
Chills	O	O			O		O				O					O
Confusion/dizziness									O	O	O	O	O		O	
Hallucinations										O	O					
Shock (low blood pressure)	O				O					O			O	O	O	O
Unable to swallow													O			
Difficulty speaking								O					O			
Dilated pupils													O			
Eye pain														O		
Photophobia											O		O			
Eyelid droop													O			
Seizure		O										O			O	

Add up the number of observable findings in the environment and from the victims. The column with the most findings gives you the probability of a weaponized chemical.

FIRST RESPONSE ACTIONS / *CBRNE Signs and Symptoms Matrix: Biological Matrix*

Section 3: Strategic Goals

- Incident Command
- Public Information
- Liaison Officer
- Terminating Officer

Operations Section Incident Command Worksheet

HAZARDOUS MATERIALS EVENT

INCIDENT #:_____ DATE:_____
LOCATION:_____

CONFINED SPACE OPERATIONS

TIME EST:_____ Ops Period: From_____
To_____

TERMINATE:_____

Combined Special Operations (Clan Lab, EOD, SWAT)

Operational Objectives

1. Site Management and Control

Approach and Position
- ☐ Approach from an uphill/upwind direction and notify units
- ☐ Conditions found through observation (visible clues)

Site Management
- ☐ Establish command
- ☐ Isolate immediate area (1,000 ft for flammable/toxic releases)
- ☐ Establish Hazmat Branch
- ☐ Request additional resources (see Staging Areas)
- ☐ Establish resource list (see Resource)

Staging Areas
- ☐ Identify staging area
- ☐ Assign an officer for Level II staging maintenance

Public Protection
- ☐ Evacuation (establish zones)
- ☐ Protect in place
- ☐ Public information

2. Identify the Problem

Survey the Incident
- ☐ Surrounding conditions and exposures
- ☐ Hazard behavior

Defining Criteria
- ☐ Occupancy and location
- ☐ Container shapes and sizes
- ☐ Markings and colors
- ☐ Labels and placards
- ☐ Shipping papers and facility documents
- ☐ Monitoring ana detection equipment

Identification
- ☐ Chemical
- ☐ Chemical and physical properties

Classification
- ☐ Behavioral event
- ☐ Damage assessment
- ☐ Monitoring and detection

Verification
- ☐ Evaluate and correlate classification
- ☐ Verify conditions and factors presented

3. Hazard and Risk Evaluation

Hazard Assessment
- ☐ Identify the hazard
- ☐ Behavioral event
  - ☐ Damage assessment

- ☐ Hazard conditions (type of container, quantity, and harmful effects (see Identification).
- ☐ Evacuation – ID relocation centers and evacuation routes/transportation.
- ☐ ID media briefing Time:_____

- ☐ Exposures
  - ☐ Population:_____
  - ☐ Chemical:_____
- ☐ Visual clues
  - ☐ Clouds/plumes
  - ☐ Fire/smoke
- ☐ Recognize identifiable clues and gather information.

- ☐ Gather the clues from the recognition and identification clues: Define the criteria of the incident
- ☐ Use this information for evacuation or protect-in-place criteria
- ☐ Identify the potential Chemical(s)

- ☐ Type of container–state of matter
- ☐ DOT Markings–DOT-ERG
- ☐ CAS#:_____
- ☐ UN #:_____
- ☐ Other:_____
- ☐ Have Information section run possibilities (see Hazard Assessment)

Behavior
- ☐ Stress
- ☐ Impingement
- ☐ Breach
- ☐ Harm
- ☐ Release
- ☐ Engulf

Damage
- ☐ Crack/score
- ☐ Gouge/dent
- ☐ Burn

- ☐ State of matter S L G
- ☐ Vapor pressure
- ☐ Flammability
- ☐ Biological
- ☐ Vapor density
- ☐ Toxicity
- ☐ Expansion ratio
- ☐ Radioactive

SAFETY PLAN

Hazard/Risk Assessment
Recommendations (from Information)
_____
_____
_____

TWA_____ STEL_____ IDLH_____
VP:_____ LEL_____ VD_____
SG_____ Other:_____

Entry Team Assessment
☐ Team A ☐ Time in:_____ ☐ Time out:_____
☐ Team B ☐ Time in:_____ ☐ Time out:_____
    ☐ Entry objective meet
    ☐ Second entry to meet objective
☐ Team A Rehabilitation
☐ Team B Rehabilitation
☐ Decon team Rehabilitation

Emergency Procedures:
_____
_____
_____
_____
_____

Safety Briefing:
☐ Chemical Behavior
☐ Incident Behavior
☐ Medical Concerns
☐ Emergency Procedures
☐ Area of Safe Refuge
☐ Communications Check

SITE PLAN

N ↑
| wind

INCIDENT ACTION PLAN

GOAL	Time
☐ Command Structure Established	
☐ Unified Command	
☐ Specialized Control	
☐ Isolation Perimeters	
☐ Hazard Control	
☐ Evacuation	
☐ Immediate Area	
☐ High Occupancy	
☐ Special Occupancy	
☐ Spill Control/Confinement	
☐ Absorption	
☐ Adsorption	
☐ Covering/Damming/Diking	
☐ Dilution	
☐ Diversion/Dispersion	
☐ Retention	
☐ Vapor Suppression/Dispersion	
☐ Leak Control/Containment	
☐ Neutralization	
☐ Over-packing	
☐ Patching and Plugging	
☐ Pressure Isolation/Reduction	
☐ Solidification/Vacuuming	
☐ Transfer	
☐ Fire Control	
☐ Extinguishing Agent	
☐ Water Supply	
☐ Reactive Chemicals	
☐ Specialized Operations	
☐ Clan Lab	
☐ EOD Support	
☐ Confined Space	
☐ SWAT Support	

INFORMATION	RECON/ENTRY	RESOURCE	DECONTAMINATION	MEDICAL
Name:_____	Name:_____	Name:_____	Name:_____	Name:_____
Radio Tac:_____	Radio Tac:_____	Radio Tac:_____	Radio Tac:_____	Radio Tac:_____
Objectives: ☐ Hazard/Risk Assessment ☐ Develop Action Plans ☐ Public Protection	Objectives: ☐ Hazard/Risk Assessment ☐ Develop Action Plans ☐ Compile Information	Objectives: ☐ Control of Supplies ☐ Tracking of Equipment ☐ Supply forecasting	Objectives: ☐ Hazard/Risk Assessment ☐ Develop Action Plans ☐ Public Protection	Objectives: ☐ Hazard/Risk Assessment ☐ Pre/Post Physicals ☐ Technical Medical

STRATEGIC GOALS / *Operations Section Incident Command Worksheet*

Hazmat Operations Public Protection: Strategic Goals and Tactical Objectives

INCIDENT ALARM #:_____ DATE:_____ TIME ESTABLISHED:_____ Ops Period

INCIDENT ADDRESS:_____ RADIO: **PIO** TIME TERMINATED:_____

STRATEGIC GOALS

	PROTECT IN PLACE		EVACUATION
	Public alert and notification (alerts in evacuation)		Public alert and notification
	Refuge area defined		Relocation facilities
	Media announcement		Media announcement

TACTICAL OBJECTIVES

	PROTECT IN PLACE		EVACUATION
	Public alert and notification		**Public alert and notification**
	Notification of occupants—emergency broadcast		Alerting of occupants
	Close all windows and doors		Fixed facility alerts—activation of sirens, or public
	Seal any obvious gaps around windows, doors, and other openings with tape		Public alert
	Turn off all HVAC		Personal notification—door to door
	Close fireplace dampers		Loudspeaker/public address system
	Turn off and cover all exhaust fans		Tone-alerted radios
	Close as many internal doors as possible and move to the shelter		Emergency Broadcast System
	Monitor local AM/FM radio or television stations for information		Scanner radios
	Refuge area defined		Television
	Type I or II structure		Sirens and alarms (emergency preparedness Sirens)

	Structure type dependent with hazmat influence	Helicopter loudspeakers
	If within 1,000 ft, consider evacuation measures	Signs—message alert systems
	Core of building	Computerized telephone notification systems
	Type III or Type IV	**Relocation facilities**
	Proximity to the event	Identified facility
	If within 1,000 ft, follow evacuation measures	Protective transportation to facility
	These building types do not offer protective function	Facility can accommodate anticipated relocated population
	Commercial structures—population high, difficult to move	Facility:_____
	Industrial structures—population low, process cannot stop	Food, toilets, showers, bedding
	Special structures—population high, difficult to move	Shelter manager
	Hospitals, nursing homes, schools, or prisons	Communication/television
	Media Announcement	
	Type of event—influences on the immediate community: _____	
	Time started, units on scene, strategies: _____	
	Special hazards and accomplishments: _____	
	Long-term influences on the community (protect in place, evacuation, road closure): _____	
	Next media announcement ($1/2$ hour before broadcast 1730, 2230 at minimum): _____	

Points to consider:
Limit media announcement to 20–30 minutes.
Start and end each announcement on time.
Fatality announcements should be through the appropriate agency.
Present in department uniform or team uniform.
Present in a professional manner.
Do not talk about mitigation techniques, only general strategies.
Do not allow the media to pull you down a negative path.

Notes:

STRATEGIC GOALS/*Hazmat Operations Public Protection: Strategic Goals and Tactical Objectives*

Liasion Officer Worksheet

INCIDENT ALARM #: _____ DATE:_____ TIME of INCIDENT: _____ TIME TERMINATED: _____

INCIDENT ADDRESS: _____ RADIO: **Liaison** OPS Period: _____

1. Site Management and Control

Approach and Position
   Conditions found through observation
Site Management
   Establish command
   Establish Hazmat Branch
   Establish resource list
Staging Areas
   Assign an officer for Level II staging maintenance
Public Protection
   Evacuation (establish zones)
   Protect in place
   Public information

2. Identify the Problem

Survey the Incident
   Surrounding conditions and exposures
   Hazard behavior
Defining Criteria
   Occupancy and location
   Container shapes and sizes
   Markings and colors
   Labels and placards
   Shipping papers and facility documents
   Monitoring and detection equipment
Identification
   Chemical:_____
   Chemical and physical properties

   BP:_____ IDLH:_____ VP:_____

   IT:_____ UEL:_____

   VD:_____ TLV(PEL):_____ LEL:_____

   SG:_____ Sol:_____

Classification
   Behavioral event
   Damage assessment
   Monitoring and detection
Verification
   Evaluate and correlate classification
   Verify conditions and factors presented

3. Hazard and Risk Evaluation

Hazard Assessment
   Identify the hazard
   Behavioral event
      Damage assessment

INTERNAL RESOURCES

INTERNAL RESOURCES

Resource	Contact Person	Tac	Physical Location

EXTERNAL RESOURCES

INTERNAL RESOURCES

Resource	Contact Person	Tac	Physical Location

Exposure Assessment
- ❒ Reference materials
- ❒ Passive analysis (visual clues)
- ❒ Active analysis (monitoring and detection)

Risk Assessment
- ❒ Quantity of material
- ❒ Containment systems

Hazard/Risk Management (see Implementing Response Objectives)

4. Select Personal Protective Clothing and Equipment

Hostile Environment _____
- ❒ Degree of hazard
- ❒ Potential outcomes (see Hazard Assessment)
- ❒ Compatibility of PPE with the chemical(s)

Task to Be Performed
- ❒ Entry and backup
- ❒ Decontamination
- ❒ Area of refuge
- ❒ Support teams

PPE Required for the Mission
- ❒ Identify the factors for suit selection
- ❒ Safety factors for operational objectives
- ❒ Approval from the commander

Capabilities of the User
- ❒ Identify operational outcomes
- ❒ Emergency entry procedures

5. Information Management and Resource Coordination

Information Management
- ❒ Interpretation of information
- ❒ Prioritization of information
- ❒ Tactical planning

Resource Coordination
- ❒ Internal
- ❒ External

6. Implement Response Objectives

Event Size-up
- ❒ Events (past, present, future)
- ❒ Behavior event
- ❒ Damage assessment

Strategic Goals
- ❒ Behavior event prediction
- ❒ Goals
  - ❒ Offensive
  - ❒ Defensive
  - ❒ Nonintervention
    - Tactical objectives
- ❒ Rescue
- ❒ Public protective actions
- ❒ Spill control
- ❒ Leak control
- ❒ Fire control
- ❒ Recovery

7. Decontamination

Decontamination Site Selected
- ❒ Site selection
- ❒ Entry and exits well marked
- ❒ Location is identified

8. Terminating the incident (see page 84)

Management Item	Time Frame—Enter hours or days									
Time ➔										
Site Position										
Site Management										
Staging Areas										
Public Protection										
Incident Forecasting										
Defining Criteria										
Chemical ID/Classification										
Hazard/Risk Assessment										
Hazard/Risk Management										
PPE Selection										
Strategic Tasks ID										
Strategic Plans										
Resource Coordination										
Decontamination										
Restoration Activity										
Termination										
Recovery										
Debriefing										

Notes:

STRATEGIC GOALS/*Liasion Officer Worksheet*

Termination Worksheet

HAZARDOUS MATERIALS EVENT

CONFINED SPACE OPERATIONS

Combined Special Operations (Clan Lab, EOD, SWAT)

Response Objectives in Progress	Termination Activities in Progress

☐ ☐ ☐ **Evaluate Isolation Area**
- ☐ Maintain control of perimeter
- ☐ Evaluate human resource to hold such perimeter
- ☐ Evaluate isolation area through hazard/risk assessment

☐ ☐ ☐ **Evaluate Behavior Event Against Current Status**
- ☐ Influence breach size
  - ☐ Contents chilled
  - ☐ Stress level limited
  - ☐ Venting activated
- ☐ Influence quantity released
  - ☐ Position of container changed
  - ☐ Pressure differential controlled
  - ☐ Breach capped off
- ☐ Influence of engulfment
  - ☐ Ignition sources controlled
  - ☐ Dikes or damming erected
  - ☐ Dilution or neutralization
- ☐ Influence of impingement
  - ☐ Shielding provided
  - ☐ Evacuation—relocation

☐ ☐ ☐ **Perform Hazard/Risk Assessment**
- ☐ Containment system evaluation
- ☐ Confinement system evaluation
- ☐ Level of hazard through active analysis
- ☐ Establish degree of hazard comparison

☐ ☐ ☐ **Identify Current PPE Requirements**
- ☐ Hazard comparison establishes:
  - ☐ Respiratory protection still required
  - ☐ Skin protection still required
- ☐ Tasks to be performed
  - ☐ Entry/backup mission
  - ☐ Current hot zone objectives
    - ☐ Behavior not controlled
    - ☐ Waiting evaluation

Scene Termination Activities
- ☐ Active analysis
  - ☐ Monitoring and detection results
    - ☐ Hot zone
    - ☐ Decontamination corridor
    - ☐ Equipment
  - ☐ Reconnaissance analysis
    - ☐ Behavior event status
    - ☐ Containment system
    - ☐ Confinement system
- ☐ Scene stabilization
  - ☐ Needs assessment for each branch
    - ☐ Information
    - ☐ Recon/entry
    - ☐ Resource
      - ☐ Internal
      - ☐ External
    - ☐ Hazmat medical
    - ☐ Decon
  - ☐ Level of service assessment
    - ☐ Equipment/supplies identified
      - ☐ Disposal
      - ☐ Decontamination
      - ☐ Re-supply
    - ☐ Functional re-deployment status
      - ☐ Adequate supplies
      - ☐ Adequate manning

- ☐ Correlated with objective status
- ☐ Decontamination level
  - ☐ Warm zone mission objectives
  - ☐ Correlated eith objective status
- ☐ Evaluation of personnel—medical status
  - ☐ Heat stress
  - ☐ Cold stress
  - ☐ Rehabilitation

Establish Resources Required and Anticipated
- ☐ Internal resources
  - ☐ Human resource status
    - ☐ Units working
    - ☐ Personnel required to maintain current objectives
    - ☐ Establish return to service plan
  - ☐ Equipment
    - ☐ Equipment required to maintain current objectives
    - ☐ Decontamination of equipment
    - ☐ Re-supply of equipment used or damaged
      - ☐ Equipment identified damaged or consumed
    - ☐ Establish return to service plan
- ☐ External resources
  - ☐ Determine level of activity required for each external resource on scene
  - ☐ Identify when each resource can be released
  - ☐ Briefing with resource assigned to section, branch, or sector

Evaluate Decontamination Status
- ☐ Containment of run-off
- ☐ Objectives not completed identified
  - ☐ Decontamination of personnel
  - ☐ Decontamination of equipment
- ☐ Medical surveillance completed

Identify Goals for Transfer
- ☐ Active analysis—hazard/risk assessment
- ☐ Evaluation of incident by safety officer/hazmat group leader
- ☐ Responsibility transfer procedures identified
- ☐ Incident debriefing
- ☐ Information-gathering responsibilities identified
- ☐ Reports from each section, branch, or sector received for report

Protect in Place
- ☐ Contact via media
  - ☐ Announce termination process
  - ☐ Time frame in which termination is complete
- ☐ Public Service Announcement

Evacuation
- ☐ Contact relocation facilities
  - ☐ Transportation schedule
  - ☐ Transportation plan
- ☐ Contact via media
  - ☐ Announce termination process
  - ☐ Time frame in which termination is complete
- ☐ Public Service Announcement

Responsibility Transfer
- ☐ Notification and evaluation of resources
  - ☐ Internal evaluation and status
  - ☐ External evaluation and status
- ☐ Identify agency/authority assuming control
- ☐ Owner/contractor briefing
  - ☐ Initial nature of the incident
  - ☐ Actions taken during incident
  - ☐ Concerns during the incident
  - ☐ Considerations of the incident
  - ☐ Chemicals involved
  - ☐ Hazard/risk assessment
    - ☐ Initial assessment
    - ☐ Current assessment

Incident Debriefing
- ☐ Identify hazards found
- ☐ Signs and symptoms of chemical exposure
- ☐ Damaged or expended equipment/supplies
- ☐ Conditions at time of transfer
- ☐ Assign information-gathering responsibilities

STRATEGIC GOALS / *Termination Worksheet*

Section 4: Tactical Goals

- Hazmat Operations
- Information
- Recon/Entry
- Hazmat Medical
- Resources
- Decontamination

Incident Action Plan

A. Survey the Incident

Surrounding conditions
    Weather
    Endangered populations
    Topography
    Exposures
Hazard behavior
    Smoke/clouds/plume
    Spill/leak/rupture
    Fire/heat/cold

C. Identification

Type of chemical involvement
Chemical and physical properties
Information evaluation

D. Classification

Damage assessment
Monitoring and detection
Logic decision tree
Chemical analysis
    Chemical classification
    Systematic chemical
    analysis

Defensive Reconnaissance

Operational Objective: To obtain information on site layout, weapon condition, physical hazards, access, and related conditions.

Operational Action: Obtained through preplans, human observation, and physical conditions of the site.

Offensive Reconnaissance

Operational Objective: To obtain incident conditions, which cannot be observed from a defensive position.

Operational Action: Monitoring, sampling, and damage assessment. Reconnaissance observation:
two in—two out.
May be combined with offensive control operations.

B. Defining Criteria

1. Occupancy and location
    Type of manufacture process
    Chemical exposures
2. Container shapes (pages 44–59)
    Size of container(s)
    Condition of the container(s)
    Type of container
3. Markings and colors (pages 60–64, 68–71)
    Color codes
    Container specification number
    Signal words
    Contents name
4. Labels and placards (pages 42–43)
    Background
    Hazard class symbol
    Hazard class/division number
    Four-digit ID number
5. Shipping papers and facility (pages 65–67) documents
    Bill of lading
    Waybill consist
    Dangerous cargo manifest
    Air bill
    MSDS
6. Monitoring and detection equipment
    Chemical family or class
7. Senses
    Observational clues

E. Verification

Match identification and classification perimeters with the seven methods of hazard categorizing.

HAZMAT GROUP STAFFING

```
                          Hazmat Group
                           Supervisor
    ┌──────────────┬───────────┼───────────┬──────────────┐
Information    Recon/Entry   Resources    Hazmat        Decon
 Research                                  Medical
```

Information Research

Tasks:
Data gathering
Coordination

Evaluate:
Hazards/risk
Public protection
Develop action plans

Recon/Entry

Tasks:
Entry/backup
Recon
Monitoring
Sampling

Evaluate:
Hazards/risk
Develop action plans

Resources

Tasks:
Control and tracking of supplies and equipment

Evaluate:
With identified hazards and projected risks, coordinates with Logistics

Hazmat Medical

Tasks:
Pre-post entry physicals
Technical medical guidance

Evaluate:
With identified hazards and projected risks, identify the medical management plan
Develop medical plan

Decon

Tasks:
Research and development of decontamination plan
Setup and evaluation of effective decon

Evaluate:
Hazards/risk
Need for additional decontamination resources for:
    ERT
    Civilians
    Equipment

HAZMAT OPERATIONS CHIEF WORKSHEET
TACTICAL OBJECTIVES

INCIDENT ALARM #: _____ DATE: _____ TIME ESTABISHED: _____ Ops Period

INCIDENT ADDRESS: _____ RADIO: **HAZMAT** TIME TERMINATED: _____ _____

Initial Objectives :	☐ Establish Hazmat Command Structure ☐ Isolation Perimeters ☐ Assistance (Level I, II, III)	☐ Law Enforcement ☐ Hazard Control Points ☐ Staging	☐ Site Manager ☐ Downwind Considerations ☐ Staging

Incident Type: ☐ Occupancy ☐ Road Transport ☐ Rail Transport ☐ Waterway	☐ Chemical: (Multiple) (Singular) ☐ Size Area: _____ ☐ State of Matter: (S) (L) (G) ☐ Isolation: _____ ☐ Evacuation: _____	Chemical (Singular): _____ Container Size: _____ Reaction Potentials: _____ Potential Stresses: _____	(Multiple) _____ _____ _____ _____ _____

	Resource		Control Actions	Weather	Public Protection Actions

Present Conditions	**Company**
	☐ Hazmat _____
	☐ Engine _____
	☐ Engine _____
	☐ Engine _____
	☐ Rescue _____
	☐ Rescue _____
	☐ Ladder _____
	☐ Ladder _____
		Radio
		_____
		_____
		_____
		_____
		_____
		_____
		_____
		_____
			☐ Site Management Setup:
			☐ Information
			☐ Recon/Entry ☐ Rescue
			                            ☐ Control
			☐ Resource
			☐ Medical
			☐ Decontamination
			☐ Initial Action Plan Developed
			from Each Management Team
				☐ Temp: _____
				☐ Wind Speed: ___
				☐ Direction: _____
				☐ Humidity: _____
				☐ Sun: _____
				☐ Clouds: _____%
				☐ Rain: _____%
				☐ Fog
				☐ Snow
					☐ Immediate Evacuation
					      ☐ Flamm/Toxic—1,000 ft.
					☐ Immediate Area
					            ☐ Evacuation
					            ☐ Protect in Place
					☐ Key Locations
					            ☐ Roads, Intersections
					☐ High-Occupancy Structures
					☐ Special Occupancy Structures

Predicted Conditions	**Company**
	☐ Hazmat _____
	☐ Engine _____
	☐ Engine _____
	☐ Engine _____
	☐ Rescue _____
	☐ Rescue _____
	☐ Ladder _____
	☐ Ladder _____
		Radio
		_____
		_____
		_____
		_____
		_____
		_____
		_____
		_____
			☐ Spill Control/Confinement
			☐ Leak Control/Containment
			        ☐ Product Transfer
			☐ Fire Control
			        ☐ Reactive Chemicals
			☐ Clandestine Drug Lab
			☐ Explosive Device Ops
			☐ Confined Space
			☐ SWAT Ops
				☐ Temp: _____
				☐ Wind Speed: ___
				☐ Direction: _____
				☐ Humidity: _____
				☐ Sun: _____
				☐ Clouds: _____%
				☐ Rain: _____%
				☐ Fog
				☐ Snow
					☐ Evacuation
					        ☐ Area: _____
					        ☐ Relocation _____
					                ☐ Services
					                ☐ Food
					                ☐ Water
					                ☐ Facilities
					☐ Protect in Place
					        ☐ Media PSA

RESOURCES

Resource	Group	Time

SITE PLAN

N
—
wind

INCIDENT ACTION PLAN

GOAL	Time
☐ Command Structure Established	
☐ Unified Command	
☐ Specialized Control	
☐ Isolation Perimeters	
☐ Hazard Control	
☐ Evacuation	
☐ Immediate Area	
☐ High Occupancy	
☐ Special Occupancy	
☐ Spill Control/Confinement	
☐ Absorption	
☐ Adsorption	
☐ Covering/Damming/Diking	
☐ Dilution	
☐ Diversion/Dispersion	
☐ Retention	
☐ Vapor Suppression/Dispersion	
☐ Leak Control/Containment	
☐ Neutralization	
☐ Over-packing	
☐ Patching and Plugging	
☐ Pressure Isolation/Reduction	
☐ Solidification/Vacuuming	
☐ Transfer	
☐ Fire Control	
☐ Extinguishing Agent	
☐ Water Supply	
☐ Reactive Chemicals	
☐ Specialized Operations	
☐ Clan Lab	
☐ EOD Support	
☐ Confined Space	
☐ SWAT Support	

INFORMATION	RECON/ENTRY	RESOURCE	DECONTAMINATION	MEDICAL
Name:_____	Name:_____	Name:_____	Name:_____	Name:_____
Radio Tac:_____	Radio Tac:_____	Radio Tac:_____	Radio Tac:_____	Radio Tac:_____
Objectives: ☐ Hazard/Risk Assessment ☐ Develop Action Plans ☐ Public Protection	Objectives: ☐ Hazard/Risk Assessment ☐ Develop Action Plans ☐ Compile Information	Objectives: ☐ Control of Supplies ☐ Tracking of Equipment ☐ Supply Forecasting	Objectives: ☐ Hazard/Risk Assessment ☐ Develop Action Plans ☐ Public Protection	Objectives: ☐ Hazard/Risk Assessment ☐ Pre/Post Physicals ☐ Technical Medical

TACTICAL GOALS / *Hazmat Operations Chief Worksheet: Tactical Objectives*

Hazard and Risk Evaluation Action Plan

Hazard and Risk Evaluation Action Plan

Each side of the action plan will require a specific set of questions. All questions and the corresponding answers represent the evaluative response. The incident determines how one must weight each question.

A. Isolate the Area

B. Deny Entry

C. Identify the Hazard
1. Occupancy and location
2. Container shapes (pages 44–59)
3. Markings and colors (pages 60–64)
4. Label and placards (pages 42–43)
5. Shipping papers and facility documents (pages 65–67)
6. Monitoring and detection equipment
7. Senses

HAZARD ASSESSMENT

State of Matter

Solid	Liquid	Gas

Melting Point	→ Boiling Point	→ Vapor Pressure

Chemical and physical influences affecting solids (particulates) and liquids	**760 mm Hg**	**Chemical and physical influences affecting gases or vapors**

RISK ASSESSMENT

Behavior Assessment (page 11)

☐ When the contents are released, where will the container or material go?
☐ What can occur to make the container or material move?
  ☐ Stress
  ☐ Breach
  ☐ Release
  ☐ Engulf
  ☐ Impinge
  ☐ Harm
☐ How will the material or the container get there?
☐ When will the event occur?
☐ What harm will the material or the container provide?

Behavior Model

Event	Cause and Effect		Correlate these factors to the state of matter	
	Stress			**ENERGY**
	Breach			**Solids**
	Release			**Liquids**
	Engulf			**Gases**
	Impinge			**Liquefied Gases**
	Harm			

92

Solids and Liquids

- ☐ Mixture
- ☐ Solution
- ☐ Slurry
- ☐ Sludge
- ☐ Cryogenic
- ☐ Color
- ☐ Temperature
- ☐ Specific gravity
- ☐ Boiling point
- ☐ Melting point
- ☐ Solubility
- ☐ Volatility

- ☐ Sublimation
- ☐ Critical temperature

Gases

- ☐ Mixture
- ☐ Vapor density
- ☐ Expansion ratio
- ☐ Vapor pressure
- ☐ Cryogenic
- ☐ Color
- ☐ Critical temperature
- ☐ Critical pressure

Toxicological Data

Least hazardous to high hazardous environments ↑

- ☐ LC_{10}
- ☐ LC_{50}
- ☐ Irreversible injury
- ☐ Immediate injury
- ☐ IDLH
- ☐ TLV-c, STEL
- ☐ PEL
- ☐ TLV-TWA
- ☐ Irritation and sensitivity reactions

Stress Event
- Thermal stress
- Mechanical stress
- Chemical stress

Breach Event
- Disintegration
- Runaway cracking
- Failure of container components
- Container punctures (see Damage Assessment)
- Container splits and tears (see Damage Assessment)

Release Event
- Detonation
- Violent rupture
- Rapid relief
- Spills or leaks

Engulfing Event
- Cloud or plume development and movement
- Form of matter/energy
- Topography, atmospheric movement

Impingement Event
- Time frame—evacuation or shelter in place

Harm Event
- Thermal
- Radiation
- Asphyxiation
- Toxicity
- Corrosivity
- Etiologic
- Mechanical

Damage Assessment

- ☐ Crack
- ☐ Score
- ☐ Gouge
- ☐ Wheel burn
- ☐ Dent
- ☐ Rail burn
- ☐ Street burn

TACTICAL GOALS / *Hazard and Risk Evaluation Action Plan*

Hazmat Information Worksheet

HAZMAT INFORMATION WORKSHEET
TACTICAL OBJECTIVES

INCIDENT ALARM #:_____ DATE:_____ TIME ESTABISHED:_____

INCIDENT ADDRESS: _____ RADIO: **INFORMATION** TIME TERMINATED: _____

Time Line		EVALUATION ACTION PLAN			
Action	**Plan Assessment**	**GOALS**			
	Information sources	**Hazard Assessment**		**Hazards Identified and Correlated**	
	Isolation/protection	Chemical data		Chemical properties	
	Mitigation	Physical data		Physical properties	
	Level of protection	Hazards found		Fire, reactivity, corrosivity, radiological	
	Decontamination	Toxicological data		Risk benefit identified	
	Medical care	**Risk Assessment**		**Probability Correlated to Chemical**	
	Manufacturer	Quantity involved		Potential quantity	
	Contractor	Containment system		Pressure/non-pressure—exposures	
	Termination	Behavior assessment		Stress event and time event	
	Medical evaluation	Damage assessment		Type of damage to the container	

Present Planning Assessment		Behavior		Future Planning Assessment	
	Preplans correlate to incident issues		Present		Preplans correlate to incident issues
	Recon information correlates to incident				Recon information correlates to incident
	Monitoring information correlates to incident		Future		Monitoring information correlates to incident
	Forecast weather conditions to incident				Forecast weather conditions to incident

94

Solid/Liquid	Flammability	Toxicity	Gas/Vapor
Melting Point:_____ Ambient Temp: _____ Solid: Milled, Particulate ☐ (surface area increased) ***Flammable—go to Flammability column*** Liquid: Size of Spill:_____ (surface area increased) Specific Gravity:_____ Water Solubility ☐ Vapor Density:_____ ***Water Reactive—go to Flammability and Reactivity column***	Flammable ☐ Combustible ☐ Flamm Range:_____ Ignition Temp:_____ Boiling Point:_____ Vapor Pressure:_____ **Reactivity** Air Reactive ☐ Water Reactive ☐ Hypergolic ☐ Polymerization ☐ SADT:_____ ☐ MMST:_____ ☐ Incompatibilities:_____ _____ _____	Referenced Values — 100% — Monitored Values (0%) Place IDLH, STEL, and TLVs in order from top to bottom. Each number should have a representative space between each entry. Left side is research values; right side is monitored values in the hot zone.	Vapor Pressure_____ Vapor Density:_____ Water Solubility_____ Expansion Ratio: ☐ ☐ Anhydrous ammonia 855:1 ☐ Argon 842:1 ☐ Carbon monoxide 680:1 ☐ Chlorine 458:1 ☐ Fluorine 981:1 ☐ Helium 745 :1 ☐ Hydrogen 850:1 ☐ Krypton 693:1 ☐ LNG 635:1 ☐ Methane 693:1 ☐ Neon 1445:1 ☐ Nitrogen 696:1 ☐ Oxygen 860:1 ☐ Propane 270:1 ☐ Xenon 559:1 ***Flammability—go to Flammability column***

TACTICAL GOALS / *Hazmat Information Worksheet*

Recon/Entry Incident Action Plan

Factors: Evaluation is based upon continuous flow of information:

- Hazard Identification
- Exposure Potential
- Degree of Risk
- Hazard/Risk Management

Defensive Reconnaissance

Operational Objective: To obtain information on site layout, weapon condition, physical hazards, access, and related conditions.
Operational Action: Obtained through preplans, human observation, and physical conditions of the site.

Offensive Reconnaissance

Operational Objective: To obtain incident conditions, which cannot be observed from a defensive position.
Operational Action: Monitoring, sampling, and damage assessment.
Reconnaissance observation:
two in—two out.
May be combined with offensive control operations.

Hazard Identification

Assessment of the vulnerability or danger the incident presents. Categorized as stress factors, behavior of the event, and damage incurred.

- ☐ Identify the Hazard
    - ☐ Survey the incident (page 6)
    - ☐ Identify defining criteria (page 6)
    - ☐ Identification process has been completed (page 7)
- ☐ Behavioral Event (Risk Assessment)
- ☐ Damage Assessment (Risk Assessment)

Exposure Potential

Identifies the influences the chemical has on its surroundings and the container from which it was released.

- ☐ Reference Materials
    - ☐ Hazard data and information
- ☐ Passive Analysis
    - ☐ Observation
    - ☐ Referencing of defining criteria with hazard data
        - ☐ Corrosivity
        - ☐ Flammability
        - ☐ Oxidizing potential
        - ☐ Oxygen potential
        - ☐ Radioactivity
- ☐ Toxicity

Hazard/Risk Management

The mitigation tool(s) applied to the incident. This must take into consideration the degree of harm, which is imposed on the entry/recon team. If the team cannot be properly protected, the tactical decision shifts from an aggressive offensive role to one that is defensive in nature.

Search/Relocation /Area at refuge established
Executing Technical Rescue
Perimeter security
Quantity of absorbent materials versus quantity of released material
Incompatibility to adsorbents
Temporary cover available and compatible
Damming – Specific Gravity (SG)
Overflow Dam SG > 1
Underflow Dam SG < 1
Diking – Limitations to a temporary measure

No available soil, area concrete or asphalt
    Frozen ground
    Equipment availability
    Human resource available

Dilution – Criteria met before action is taken as a last resort   Run off and environmental concerns

Not water reactive
    Will not generate a toxic gas
    Will not form a solid or precipitate
    Water solubility factors are high

Diversion – Placed ahead of the spill

Consider speed and quantity of material
    Greater the speed/quantity, the greater the length and angle of the barrier
    Area of involvement increases

Dispersion increases involved area
Retention – Must be employed with diversion or diking
Require resources
Vapor Dispersion/Vapor Suppression
Run off and environmental concern   May create pockets of chemical concern
Neutralization – quantity needed calculated and available
Overpacking – Type and quantity available
Patching and Plugging – Dependent on the size of opening
Pressure Isolation/reduction – referenced and equipment available – Venting, Flaring, Hot Tap, Vent and Burn
Solidification Quantity of material versus available containment
Vacuuming – Dependent on quantity, venting secured

Degree of Risk

Correlates with the potential jeopardy a population, community, or container, may have. The population may be the size of the community or it can be localized to the first responders. In either case, the hazards have to be identified through researching of the material, its physical and chemical properties, and the visual clues available.

- ☐ Active Analysis
    - ☐ Reconnaissance
        - ☐ Relate to defining criteria with hazard data
        - ☐ Correlate with chemical and physical properties
    - ☐ Monitoring and detection
        - ☐ Corrosively
            - ☐ pH paper, strips, meters
        - ☐ Flammability
            - ☐ CGI
        - ☐ Oxygen potential
            - ☐ Oxygen meter
        - ☐ Radioactivity
            - ☐ Radiation detector, dosimeters
        - ☐ Toxicity
            - ☐ Colormetrics
- ☐ Quantity of material
    - ☐ Potential additional releases
    - ☐ Dispersion characteristics
- ☐ Containment systems
    - ☐ Correlate to defining criteria
    - ☐ Stress event (behavior)
    - ☐ Physical damage (damage potentials)
        - ☐ Behavior event (stress)
        - ☐ Damage potentials of containers

TACTICAL GOALS / *Recon/Entry Incident Action Plan*

Information Management and Resource Coordination Worksheet

INCIDENT ALARM #:_____

INCIDENT ADDRESS: _____

DATE:_____

RADIO: **INFORMATION**

TIME ESTABISHED:_____

TIME TERMINATED: _____

Ops Period

_____

Current Incident Status			Hazard Zones Established	
Behavior Event		**Damage Potentials**		
	Breach		Crack/Gouge	
	Release		Score/Dent	
	Engulf		Burns	
	Impingement		Vent	
Defining Criteria				
	Occupancy Type			
	Container Type			
	Markings			
	Labels/Placards			
	Shipping Papers/Facility Documents		* Place estimated distances of all zones and dimensions	

Branch Notification and Information Distribution		Current Weather Conditions:	Forecasted Weather Conditions:
	Hazmat Branch Officer	Temperature:_____	Temperature:_____
	Hazmat Safety Officer	Wind Speed/Direction:_____	Wind Speed/Direction:_____
	Information Officer	Humidity/Dew Point:_____	Humidity/Dew Point:_____
	Entry Officer	Barometric Pressure:_____	Barometric Pressure:_____
	Resource Officer	Sun/Clouds % Cover_____	Sun/Clouds % Cover_____
	Decontamination Officer	Rain/Fog/Snow/Sleet	Rain/Fog/Snow/Sleet

Chemical(s) Involved:

Technical Information Sources Referenced (minimum of three separate references)

General Reference	Page #	General Reference	Page #
DOT Guidebook		CHRIS Manual	
NFPA Hazmat Manual		Handbook of Toxic and Hazardous Chemical Carcinogens	
AAR Manual		Geniums Handbook of Safety and Environment	
AAR Emergency Action Plans		The Firefighter's Handbook of Hazardous Materials	
Sax Handbook to Industrial Chemicals		Hawley's Condensed Chemical Dictionary	
Merck Index		NIOSH Pocket Guide to Chemical Hazards	

Specialized Reference	Page #	Specialized Reference	Page #
Farm Chemicals Handbook		Recognition and Management of Pesticides	
Guidelines for Selection of PPE		GATX Tank Car Manual	
Handbook of Compressed Gases		Handbook of Reactive Chemicals	
Emergency Medical Care for Hazmat Exposure		Clinical Toxicology of Commercial Products	

Internet Reference	URL	Internet Reference	URL

Computer Databases	Search String	Computer Database	Search String
TOMES		Sax Handbook to Industrial	
Merck Manual		Hazardous Desk Reference	

Emergency Medical Information

☐ Correlate with Recon/Entry Incident Action Plan worksheet (pages 96–97) and Medical Worksheet (page 101)
☐ Correlate with Hazmat Information Worksheet (page 94)
☐ Common route of entry:_____
☐ Target organs affected:_____
☐ Signs and symptoms:_____
_____

Decontamination Information

☐ Correlate with PPE objectives (page 14)
☐ Incompatibilities:_____
☐ State of matter:_____
☐ Decontamination solutions required:_____
☐ Protective garments:_____
☐ Respiratory protection:_____
☐ Containment required – (Y) (N) Notification:_____
_____

TACTICAL GOALS / *Information Management and Resource Coordination Worksheet*

Personal Protective Equipment Worksheet

INCIDENT ALARM #:_____ DATE:_____ TIME ESTABISHED:_____ Ops Period

INCIDENT ADDRESS: _____ RADIO: **ENTRY SUPPORT** TIME TERMINATED: _____ _____

Environment		Tasks		PPE Required	
Action plan developed		Entry personnel		Medical evaluation performed	
Type of mission _____		Hot zone objectives identified		Suit compatibility double-checked	
Information advised of:		Suit limitations		Duration of the operation briefing	
Detail of outcomes		Objectives match suit limitations		Degree of decon required	
Dispersion patterns		Decontamination		Backup same as entry	
Concentrations expected		Warm zone Identified		Safety objectives:	
Potential behavior		Suit identified and communicated		Hand signals and communication	
Suit compatibility		Support teams—backup		Medical monitoring/rehab	
Level of respiratory protection		Objectives briefing		Verify air supply/protection factors	
Suit/glove compatibility		Placement of backup		Approval of protection from IC	
PRESENT Potential harm		Area of refuge identified		Suit/air/level of protection checked	
Suit compatibility		Emergency procedures briefing		Visual check of entry suit	
Decon setup		Medical team(s) in place		Final communications check	
FUTURE Potential harm		Support of area of refuge		Backup in suitable protection level	
Suit compatibility		Level of protection required		Secondary backup	
Resources (discuss with IC)		Resources (discuss with IC)		Resources (discuss with IC)	

Potential Chemicals	
Chemical	
Chemical	

MEDICAL WORKSHEET

| INCIDENT ALARM #:_____ | DATE:_____ | TIME ESTABLISHED:_____ | Ops Period |
| INCIDENT ADDRESS: _____ | RADIO: **MEDICAL** | TIME TERMINATED: _____ | _____ |

ENTRY TEAM SUPPORT

TEAM				TASKS
A	B	C	Time	(See Entry and Medical Status Worksheet)
				Medical monitoring of entry and backup personnel
				All personnel items removed, tagged, and secured
				Suit selection double-checked with information
				Protective clothing: _____
				Visual check of entry suit
				All zippers and closures properly secured
				No obvious suit damage
				Communication check (channel: _____)
				Respiratory protection: _____
				Facepiece seal ensured
				Air pressure verified
				Gloves: _____
				Gloving, overgloving, double-glove verified
				Boots: _____
				Footwear verified as appropriate
				Entry officer notified that teams are ready
Chemical				

Initial Entry Physical

Suit Numbers

TEAM A	TEAM B	TEAM C

Respiratory Protection Numbers

TEAM A	TEAM B	TEAM C

TACTICAL GOALS / *Medical Worksheet*

Selection of PPE and Medical Action Plan

Variables from Hazard Risk Evaluation:
- Solid—melting point
- Liquid—boiling point
- Gas—vapor pressure
- Influences from the environment
- Influences from the behavior of the chemical

Suit Selection:
- Identification of chemical
  - Chemical properties and state of matter
  - Toxicological values scaled
- Permeation values for the chemical against fabric/suits
- Incompatibilities of chemical
- Correlation of incompatibilities with protection variables (permeation)

Offensive Operations
- Entry procedures
  - Mission objectives
    - Suit/glove/boot compatible with objectives

Defensive operations
- Protective procedures
  - Immediate area isolation with appropriate skin/respiratory protection
- Protect in place/evacuation (page 3)

Solids and Liquids
- ☐ Mixture
- ☐ Solution
- ☐ Slurry
- ☐ Sludge
- ☐ Cryogenic
- ☐ Color
- ☐ Temperature
- ☐ Specific gravity
- ☐ Boiling point
- ☐ Melting point
- ☐ Solubility
- ☐ Volatility

☐ Sublimation
☐ Critical temperature

Gases
- ☐ Mixture
- ☐ Vapor density
- ☐ Expansion ratio
- ☐ Vapor pressure
- ☐ Cryogenic
- ☐ Color
- ☐ Critical temperature
- ☐ Critical pressure

Toxicological Data

Least hazardous to high-hazardous environment

- ☐ LC_{10}
- ☐ LC_{50}
- ☐ Irreversible injury
- ☐ Immediate injury
- ☐ IDLH
- ☐ TLV-c, STEL
- ☐ PEL
- ☐ TLV-TWA
- ☐ Irritation and sensitivity reactions

Decontamination Operations
- Entry team decontamination
- Mass decontamination

Type of Operation

- Type of mission versus degree of hazard
    - Exposures/harm
        - Human exposures (rescue potential)
        - Lethality of chemical (nonrescue)
        - Property damage potential
        - Environmental potential
    - Scene orientation
        - Defensive
        - Offensive
        - Nonintervention
    - Identify the degree of potential harm
        - Dependent on level of engagement (work area) versus possible level of protection
        - Entry and backup team mission level
- Compatibility
    - Degradation—Temperature dependent measured at 70–75°F
    - Penetration—Suit testing and inspection schedules

Permeation—Breakthrough time (Most charts describe this as breakthrough time and permeation rates.)

- Signs/symptoms of chemical threat
    - Early signs
    - Late signs
- Technical advice
    - Posion Control
    - Hospital notification

☐ Action plan developed (page 78)
  ☐ Type of mission versus degree of hazard
    ☐ Identify the response objectives (page 90)
  ☐ Detail possible chemical/physical outcomes based on
    ☐ Referenced material
    ☐ Environmental concerns
    ☐ Resource capability
☐ Potential outcomes
  ☐ Dispersion patterns
  ☐ Size, shape, and concentrations associated with release
  ☐ Correlate with defined criteria and potential behavior of the container
☐ Compatibility of PPE with the chemical(s)
  ☐ Identify the suit compatibility
  ☐ Match suit and glove compatibility
  ☐ Identify the level of respiratory protection
☐ Entry and backup mission
  ☐ Hot zone mission objectives identified
  ☐ Based on suit limitations/advantages
  ☐ Objectives correlated with mission and availability of suits
☐ Decontamination team mission level
  ☐ Warm zone mission objectives identified
  ☐ Based on suit limitations/advantages
  ☐ Objectives correlated with mission and availability of suits
☐ Area of refuge mission
  ☐ Warm zone safety mission identified

TACTICAL GOALS/*Selection of PPE and Medical Action Plan*

Medical Status Worksheet

INCIDENT ALARM #:_____ DATE:_____ TIME ESTABISHED:_____ Ops Period

INCIDENT ADDRESS: _____ RADIO: **MEDICAL** TIME TERMINATED: _____ _____

ENTRY TEAM _____

Personnel: Name:_____

Initial Entry Physical:
☐ Blood Pressure :_____/_____
☐ Pulse :_____
☐ Respiration :_____
☐ Weight :_____
    ☐ Hydration
    ☐ Temp:

Suit:_____
Suit #:_____
Air Pressure:_____
Tank #: _____
Time In:_____

Initial Exit Physical:
☐ Blood Pressure :_____/_____
☐ Pulse :_____
☐ Respiration :_____
☐ Weight :_____
    ☐ Hydration
    ☐ Temp:

Area Entered:_____

Mission/Task:_____
Air Pressure:_____
Time Out:_____

Rehabilitation: Time In:_____
           Time Out:_____

Minimum of 30 minutes gauged against weight.

Second Entry Physical:
☐ Blood Pressure :_____/_____
☐ Pulse :_____
☐ Respiration :_____
☐ Weight :_____
    ☐ Hydration
    ☐ Temp:

Suit:_____
Suit #:_____
Air Pressure:_____
Tank #: _____
Time In:_____

Second Exit Physical:
☐ Blood Pressure :_____/_____
☐ Pulse :_____
☐ Respiration :_____
☐ Weight :_____
    ☐ Hydration
    ☐ Temp:

Area Entered:_____

Mission/Task:_____
Air Pressure:_____
Time Out:_____

Rehabilitation: Time In:_____
           Time Out:_____

Minimum of 30 minutes gauged against weight

Chemical Name:

Hazard Assessment:

Mission Objective:

Personnel: Name:_____

Initial Entry Physical:
☐ Blood Pressure :_____/_____
☐ Pulse :_____
☐ Respiration :_____
☐ Weight :_____
    ☐ Hydration
    ☐ Temp:

Suit:_____
Suit #:_____
Air Pressure:_____
Tank #: _____
Time In:_____

Initial Exit Physical:
☐ Blood Pressure :_____/_____
☐ Pulse :_____
☐ Respiration :_____
☐ Weight :_____
    ☐ Hydration
    ☐ Temp:

Area Entered:_____

Mission/Task:_____
Air Pressure:_____
Time Out:_____

Rehabilitation: Time In:_____
           Time Out:_____

Minimum of 30 minutes gauged against weight.

Second Entry Physical:
☐ Blood Pressure :_____/_____
☐ Pulse :_____
☐ Respiration :_____
☐ Weight :_____
    ☐ Hydration
    ☐ Temp:

Suit:_____
Suit #:_____
Air Pressure:_____
Tank #: _____
Time In:_____

Second Exit Physical:
☐ Blood Pressure :_____/_____
☐ Pulse :_____
☐ Respiration :_____
☐ Weight :_____
    ☐ Hydration
    ☐ Temp:

Area Entered:_____

Mission/Task:_____
Air Pressure:_____
Time Out:_____

Rehabilitation: Time In:_____
           Time Out:_____

Minimum of 30 minutes gauged against weight

HAZARDOUS MATERIALS RESOURCES WORKSHEET

INTERNAL RESOURCES

Resource	Radio Designation	Radio Tac	Objective/Assignment	Time In	Time Out
Fire					
Law Enforcement					
Public Works					

TACTICAL GOALS/*Hazardous Materials Resources Worksheet*

Hazardous Materials Resources Worksheet

EXTERNAL RESOURCES

Resource Local/State Agency	Contact Number	Contact Person	ETA	Objective/Assignment	Time In	Time Out
Health Department						
Water Commission						
County EPA						
Contractors:						
Federal Resource						
U.S. Coast Guard						
U.S. EPA						
National Response Center						
CHEMTREC						
FBI						

Resource Tracking Equipment Item	Quantity Available	Quantity Used	Net Amount	Supplier	Contact Number	Contact Person
PROTECTIVE ENSEMBLES						
SPILL AND LEAK CONTROL						
DECON EQUIPMENT						
OTHER—MISCELLANEOUS						

TACTICAL GOALS / *Hazardous Materials Resources Worksheet*

Resource Management Action Plan

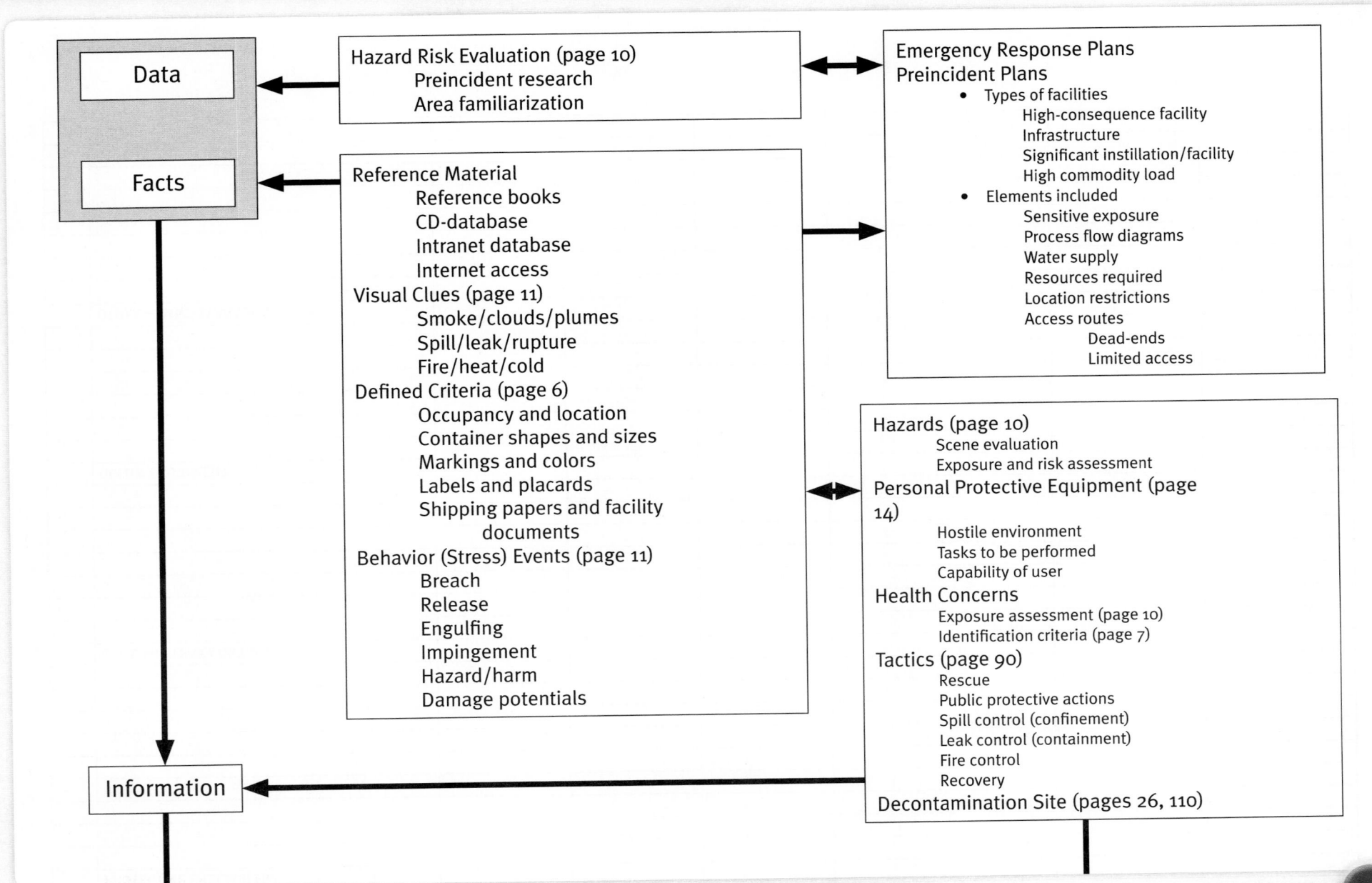

Data

Facts

Hazard Risk Evaluation (page 10)
- Preincident research
- Area familiarization

Reference Material
- Reference books
- CD-database
- Intranet database
- Internet access

Visual Clues (page 11)
- Smoke/clouds/plumes
- Spill/leak/rupture
- Fire/heat/cold

Defined Criteria (page 6)
- Occupancy and location
- Container shapes and sizes
- Markings and colors
- Labels and placards
- Shipping papers and facility documents

Behavior (Stress) Events (page 11)
- Breach
- Release
- Engulfing
- Impingement
- Hazard/harm
- Damage potentials

Emergency Response Plans
Preincident Plans
- Types of facilities
  - High-consequence facility
  - Infrastructure
  - Significant instillation/facility
  - High commodity load
- Elements included
  - Sensitive exposure
  - Process flow diagrams
  - Water supply
  - Resources required
  - Location restrictions
  - Access routes
    - Dead-ends
    - Limited access

Hazards (page 10)
- Scene evaluation
- Exposure and risk assessment

Personal Protective Equipment (page 14)
- Hostile environment
- Tasks to be performed
- Capability of user

Health Concerns
- Exposure assessment (page 10)
- Identification criteria (page 7)

Tactics (page 90)
- Rescue
- Public protective actions
- Spill control (confinement)
- Leak control (containment)
- Fire control
- Recovery

Decontamination Site (pages 26, 110)

Information

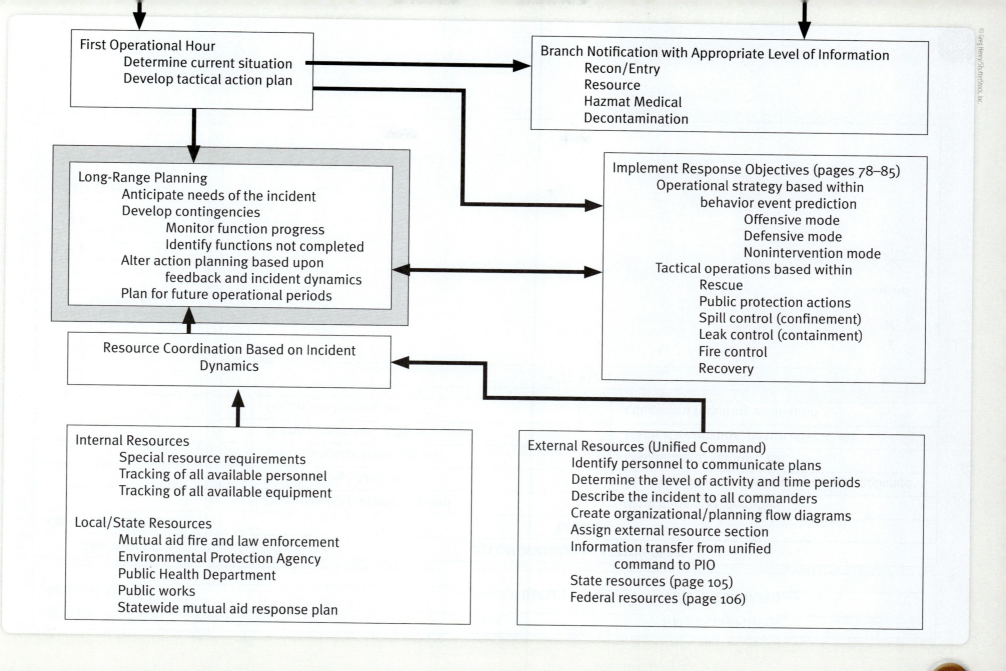

TACTICAL GOALS / *Resource Management Action Plan*

Decontamination Officer Worksheet

INCIDENT ALARM #:_____

INCIDENT ADDRESS: _____

DATE:_____

RADIO: **DECON**

TIME ESTABISHED:_____

TIME TERMINATED: _____

RESOURCES

CO	ASSIGNMENT

CHEMICAL

_____

DOT #:_____

CAS #: _____

Health Concerns:

_____

_____

DECONTAMINATION ACTION PLAN

GOALS

	Decontamination site selected		Work perimeter identified
	Resources established		Solutions identified—water source established
	Decontamination site set up		Containment, stations, Etc.
	Decon personnel—EMS evaluation		Medical evaluation, health effects, etc.
	Decon site termination		Equipment requiring evaluation

SITE PLAN

N

wind

Decontamination Tactical Overview

Decontamination Site Selected

- Decon station well marked
- ☐ Topography and wind considered
    - ☐ Area sloped toward entrance
    - ☐ Based on wind and ground contours
- ☐ Entry and exit established
- ☐ Located in the warm zone
- Water source established
    - ☐ Run-off considered
    - ☐ Decon solution identified
- Sufficient disposal containers
- ☐ Equipment
- ☐ Evidence
- Area of refuge identified

Resources Established

- PPE identified
    - ☐ Protective clothing
    - ☐ Outer garment
    - ☐ Gloves and boots
    - ☐ Respiratory protection
        - ☐ Extra SCBA, air bottles, filters
- Decon equipment
    - ☐ Water source
    - ☐ Supportive equipment
        - ☐ Containment pools
        - ☐ Water hoses and sprayers
        - ☐ Pails and brushes for each station
        - ☐ Decon solution mixed
- Sufficient disposal containers
    - ☐ Equipment disposal
    - ☐ Evidence containment

Decon Site Set Up

- Chemical effects on equipment
    - ☐ Reactivity to water
    - ☐ Reactivity to decon solutions
- Personal showering facilities
    - ☐ On site/Off site
- Decon solution containment
    - ☐ Containment
        - ☐ Over-pack or basins for containment
    - ☐ Permitted into sewers
- Equipment is in position
    - ☐ Decon team in position
    - ☐ Decon procedures
        - ☐ Emergency decon identified
    - ☐ Area of refuge identified
    - ☐ Stations identified
        - ☐ Clean side versus dirty side
        - ☐ Entrance point
        - ☐ Technical decon
        - ☐ SCBA removal
        - ☐ Removal and isolation of PPE
        - ☐ Removal of personal clothing
        - ☐ Body wash
        - ☐ Dry off and don clean clothing
        - ☐ Medical evaluation

Decon Personnel—Evaluation

- Health effects of exposure
    - ☐ Signs and symptoms of chemical
    - ☐ EMS preparation for response personnel
    - ☐ Emergency decon and responder assessment
- Hospital notification
    - ☐ Personnel decontaminated prior to transport
    - ☐ Victims decontaminated prior to transport
    - ☐ Number of potential patients
- Medical surveillance of response personnel
    - ☐ Decon Steps 8 and 9, medical evaluation
    - ☐ Injured: advanced medical evaluation
        - ☐ Removed by
        - ☐ Minimal treatment in contaminated areas
        - ☐ Maintain basic treatment within decon
        - ☐ Remove and isolate contaminated clothing

TACTICAL GOALS / *Decontamination Officer Worksheet*

Decontamination Incident Action Plan

A. Decontamination need has been identified (see Hazmat Information Worksheet, page 94)

B. Decontamination strategies are implemented (see Decontamination Officer Worksheet, page 110)

C. Health and safety issues examined
   Entry team briefed on preventive steps
   1. Minimize contact (surface contamination)
   2. Identified zone of refuge
   3. Disposable undergarment
   4. Limited use or disposable if appropriate
   5. Equipment protected
D. Prevention of contamination
   Permeation contamination
   1. Contact time
   2. Temperature (ambient hot or cold)
   3. Physical state of matter (solid, liquid, gas)

Special Operations Considerations
1. Emergency decontamination procedures
2. Establish an area of refuge
3. Maintain clean side/dirty side
4. Consider all types of entry
   a. Rescue and reconnaissance
   b. Mitigation
   c. Law enforcement (SWAT)
   d. Crime scene investigation
5. Evidence decontamination
   a. Requires a chain of custody process
6. Perpetrator/prisoner decontamination
7. Decontamination security
8. Access control points

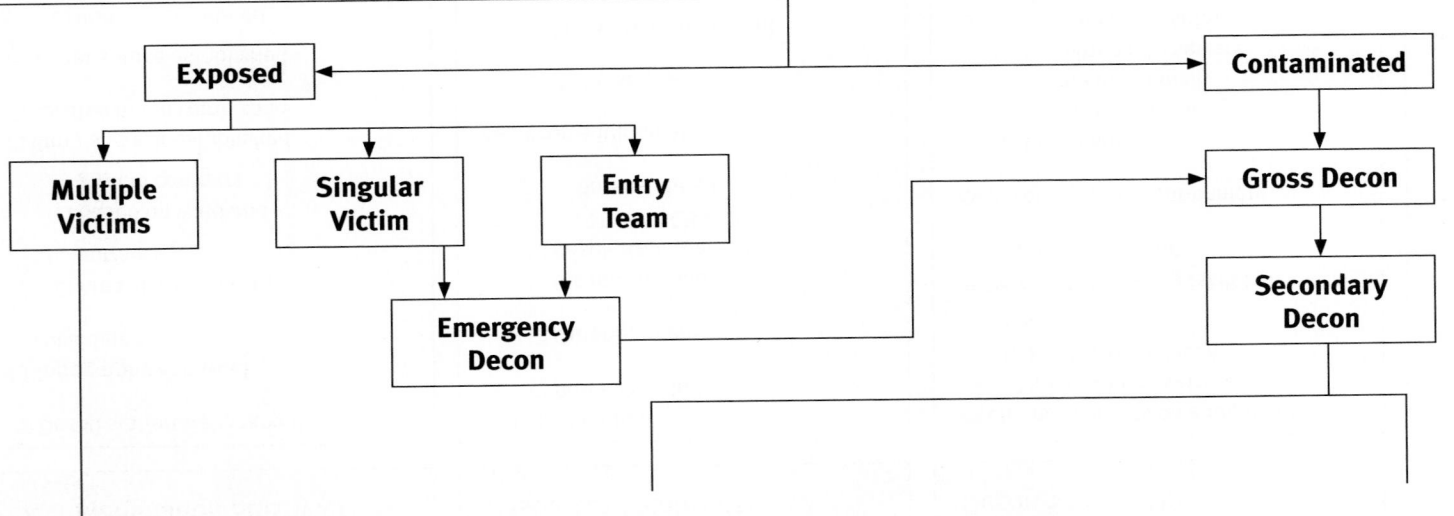

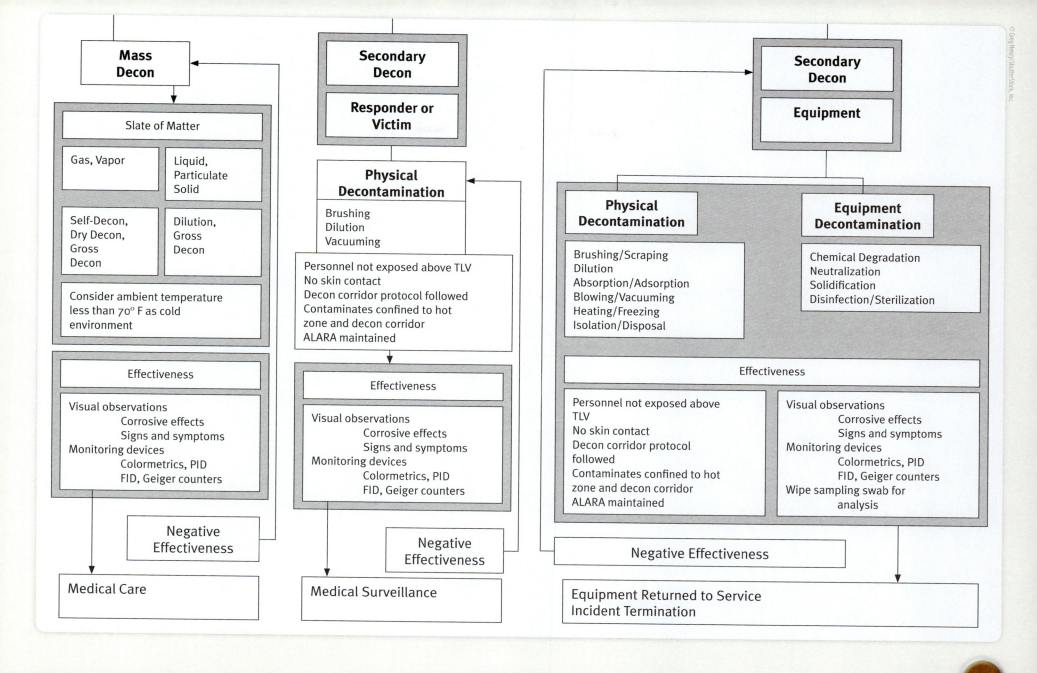

TACTICAL GOALS / *Decontamination Incident Action Plan*

Top 35 Chemicals with CBRNE

SECTION 5

- Chemical Scan Sheets

TOP 35 CHEMICALS

Chemical Scan Sheets

Chemical Name	Synonym	CBRNE	DOT #	CAS #	RETCS#
Acetic acid	Ethanoic acid, ethylic acid, AAC		2790	64-19-7	AF1225000
			2789		
Acetone	2-Propanone, dimethyl ketone		1090	67-64-1	AL3150000
Acrylic acid	2-Propenoic acid, vinylformic acid		2218	79-10-7	AS4375000
Ammonia	Aqueous ammonia, anhydrous ammonia		1005	7664-41-7	BO0875000
			2672		
			2073		
Arsine	Arsenic trihydride	SA	2188	7784-42-1	CG6475000
Benzene	Benzol, cyclohexatriene		1114	71-43-2	CY1400000
Butadiene	1,3-Butadiene, vinylethylene		1010	106-99-0	EI9275000
Benzilate	1-Azabicyclo-2,2,2-octan-3-ol	BZ (QNB)	2810	6581-06-2	VD6300000
Carbon monoxide	Monoxide, carbon oxide		1016	630-08-0	FG3500000
Chlorine	Chlogas, CLX	CL	1017	7782-50-5	FO2100000
Cyanogen chloride	Chlorine cyanide, chlorocyanide	CK	1589	506-77-4	GT2275000
Ethyl mercaptan	Ethanethiol, ethyl sulfhydrate		2363	75-08-1	KI9625000
Ethyldichloroarsine	Dichloro(ethyl)arsine	ED	1892	598-14-1	CH3500000
Ethylene	Ethene, olefiant gas, ETL, Elayl		1083	74-85-1	KU5340000
			3138		
			1962		
Formaldehyde	Methyl aldehyde, methanal		1198	50-00-0	LP8925000
			2209		
Gasoline	Petrol, motor fuel, natural gasoline		1203	8006-61-9	LX3300000
Hydrogen chloride	Hydrochloric acid, muriatic acid		1050	7647-01-0	MW4025000
			1789		
Hydrogen cyanide	Prussic acid, hydrocyanic acid	AC	1051	74-90-8	MW6825000
	Formonitrile		1613		
Hydrogen fluoride	Hydrofluoric acid gas, anhydrous hydrofluoric acid		1052	7664-39-3	MW7875000
			1790		
Hydrogen peroxide	Hydroperoxide, hydrogen dioxide		2984	7722-84-1	MX0900000
			2014		
			2015		
Hydrogen sulfide	Sulfureted hydrogen, stink damp		1053		MX1225000
Lewsite	Dichloro-(2-chlorovinyl) arsine	L	2810	541-25-3	CH2975000
Methane	Marsh gas, methyl hydride, LNG		1972	74-82-8	PA1490000
Naphtha	High-solvent naptha, coal tar		1256	8030-30-6	DE3030000
			1255		
			2553		
			1268		

Chemical Name	Synonym	CBRNE	DOT #	CAS #	RETCS
Nitric acid	Hydrogen nitrate, aqua fortis		1760	7697-37-2	QU5775000
			2031		
			2032		
Nitrogen mustard	Ethylbis(2-chloroethyl) amine	HN (HN-1)	2810	538-07-8	YE1225000
	Chloramine	HN (HN-2)		51-75-2	IA1750000
	Tri-(2-chloroethyl) amine	HN (HN-3)		555-77-1	YE2625000
Oleum	Sulfuric acid, mixture with sulfur trioxide		1831	8014-95-7	WS5605000
Phenol			1671	108-95-2	SJ3325000
			2312		
			2821		
Phosgene	Carbonyl chloride	CG	1076	75-44-5	SY5600000
Phosgene oxime	Dichloroformoxime	CX	2811	1794-86-1	
Phosphoric acid	Orthophosphoric acid		1805	7664-38-2	TB6300000
Potassium hydroxide	Caustic potash, potassium hydrate		1813	1310-58-3	TT210000
			1814		
Potassium permangenate	Cairox, chameleon mineral		1490	7722-64-7	SD6475000
Propane	Dimethylmethane, petroleum gas		1075	74-98-6	TX2275000
	Propyl hydride, LPG, PRP		1978		
Pyridine	Azine, azabenzene		1282	110-86-1	UR8400000
Sarin	Isopropyl methylfluorophosphate	GB	2810	107-44-8	TA8400000
Sodium hydroxide	Caustic soda, soda lye, lye		1823	1310-73-2	WB4900000
	Sodium hydrate		1824		
Soman	Methyl pinacolyl phosphonofluoridate	GD	2810	96-64-0	TA8750000
Styrene	Cinnamene, ethylbenzene		2055	100-42-5	WL3675000
Sulfur dioxide	Sulfurous anhydride, sulfur oxide		1079	9/5/46	WS4550000
Sulfur mustard	Yellow cross	HD	2810	505-60-2	WQ0900000
Sulfuric acid	Oil of vitrol		1830	7664-93-9	WS5600000
			1831		
			1832		
Tabun	Dimethylamidoethyoxyphosphoryl cyanide	GA	2810	77-81-6	TB4550000
Tolulene	Methylbenzene, toluol, phenyl methane		1294	108-88-3	XS5250000
Vinyl chloride	Chloroethylene, chloroethene		1086	75-01-4	KU9625000
Vx	Ethyl-S-dimethylaminoethyl methylphosphonothiolate VX		2810	50782-69-9	TB1090000
Xylene	Dimethylbenzene		1307	95-47-6	ZE2450000
	m-Xylol			108-38-3	ZE2275000
	p-Xylol			106-42-3	ZE2625000

TOP 35 CHEMICALS / *Chemical Scan Sheets*

ACETIC ACID
CH$_3$COOH

UN # 2790 (10–80%)
       2789 (>80%)
CAS # 64-19-7
RTECS # AF1225000

Poison
Flammable
Corrosive

GUIDE # 153 (10–80%)
     132 (>80%)

DOT-ERG Suggestions:
Fire Isolation: 800 meters in all directions

Spills: Small: 50–100 meters isolation

Large: 800 meters isolation

CHEMISTRY

Vapor Pressure	16 mm Hg
Vapor Density	3.45
MW	60.05
UEL	16%
LEL	4%
Flash Point	109°F
Ignition Temp.	705.2°F
Specific Gravity	0.9276
Solubility	Slightly
Boiling Point	218°F
Ion Potential	10.66 eV

PPE

Depending on the concentration and quantity of the spilled material, wear the appropriate level of protection to prevent the possibility of skin contact with liquids of greater than 50% or prolonged environment in concentrations of 10–49%. Skin and eye protection should be worn.

>10 to <50 ppm	SAR
50 ppm and greater	SCBA

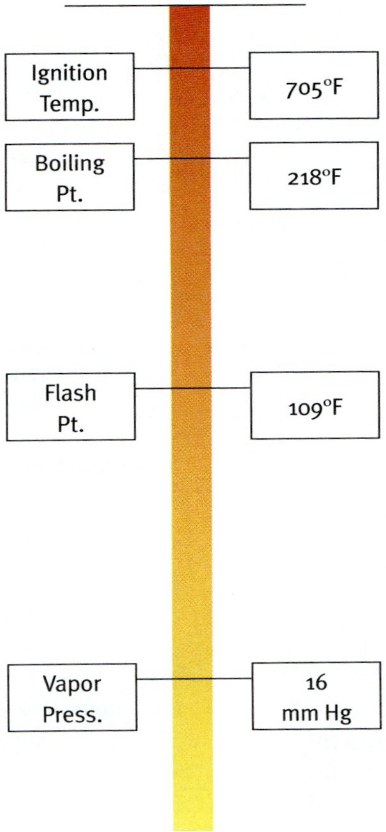

Ignition Temp.	705°F
Boiling Pt.	218°F
Flash Pt.	109°F
Vapor Press.	16 mm Hg

TOXICOLOGY

LEL	40,000 ppm	4%
LEL 10%	4,000 ppm	0.4%
LC$_{50}$	1,000 ppm	0.1%
IDLH	50 ppm	0.005%
Odor	0.21 ppm	0.000021%
STEL	15 ppm	0.0015%
PEL	10 ppm	0.001%
TLV-TWA	10 ppm	0.001%

DECONTAMINATION

Wash immediately with soap and water, removing contaminated clothing. Constant irrigation to ensure removal of the contaminate from soft tissue and mucous membranes.

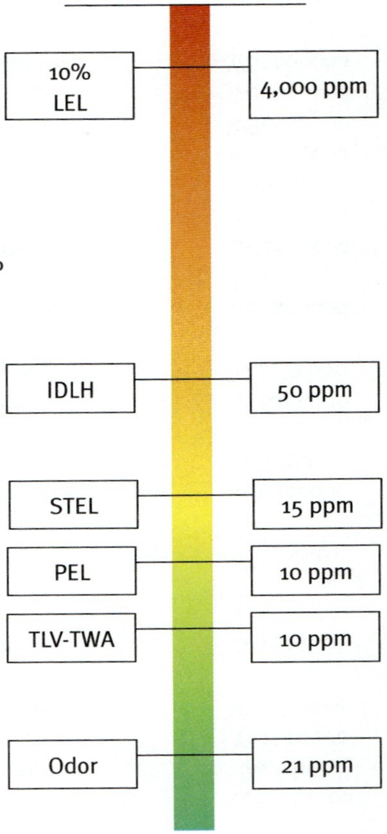

10% LEL	4,000 ppm
IDLH	50 ppm
STEL	15 ppm
PEL	10 ppm
TLV-TWA	10 ppm
Odor	21 ppm

ACETONE $(CH_3)_2CO$

UN # 1090
CAS # 67-64-1
RTECS # AL3150000

Flammable

GUIDE # 127

DOT-ERG Suggestions:
Fire Isolation: 800 meters in all directions
Spills: Small: 25–50 meters isolation
Large: 300 meters isolation

CHEMISTRY

Vapor Pressure	180 mm Hg
Vapor Density	2
MW	58.09
UEL	12.8%
LEL	2.5%
Flash Point	−4°F
Ignition Temp.	869°F
Specific Gravity	0.7899
Solubility	Miscible
Boiling Point	133°F
Ion Potential	9.69 eV

PPE

Depending on the quantity of the spilled material, wear the appropriate level of protection to prevent the possibility of skin contact and respiratory tract involvement, by wearing protective ensemble with self-contained breathing apparatus.

>1,000 to <2,500 ppm   SAR
2,500 ppm and greater   SCBA

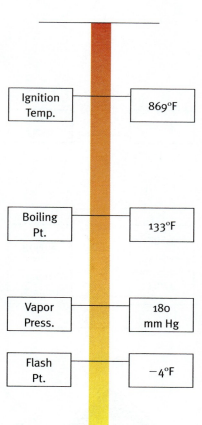

TOXICOLOGY

LEL	25,000 ppm	2.5%
LEL 10%	2,500 ppm	0.25%
LC_{50}		
IDLH	2,500 ppm	0.25%
Odor	13 ppm	0.00013%
STEL	1000 ppm	0.1%
PEL	1000 ppm	0.1%
TLV-TWA	750 ppm	0.075%

DECONTAMINATION

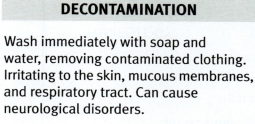

Wash immediately with soap and water, removing contaminated clothing. Irritating to the skin, mucous membranes, and respiratory tract. Can cause neurological disorders.

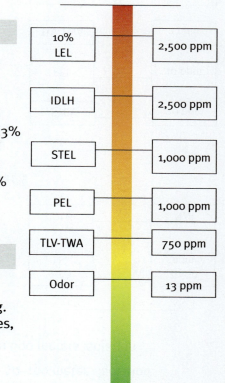

ACRYLIC ACID
$CH_2=CHOOH$

UN # 2218 Inhibited—
Polymerization
CAS # 79-10-7
RTECS # AS4375000

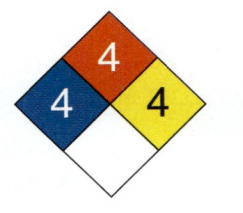  

Polymerization
Flammable
Corrosive

GUIDE # 132P

DOT-ERG Suggestions:
Fire Isolation: 800 meters in all directions

Spills: Small: 50–100 meters isolation

Large: 800 meters isolation

CHEMISTRY

Vapor Pressure	4 mm Hg
Vapor Density	2.5
MW	72.06
UEL	20.2%
LEL	2.4%
Flash Point	114°F
Ignition Temp.	734°F
Specific Gravity	1.0497
Solubility	Soluble
Boiling Point	285.8°F
Ion Potential	? eV

TOXICOLOGY

LEL	24,000 ppm	2.4%
LEL 10%	2,400 ppm	0.24%
LC_{50}	4,000 ppm	0.4%
IDLH		
Odor	1.04 ppm	0.000021%
STEL	20 ppm	0.002%
PEL	10 ppm	0.001%
TLV-TWA	2 ppm	0.0002%

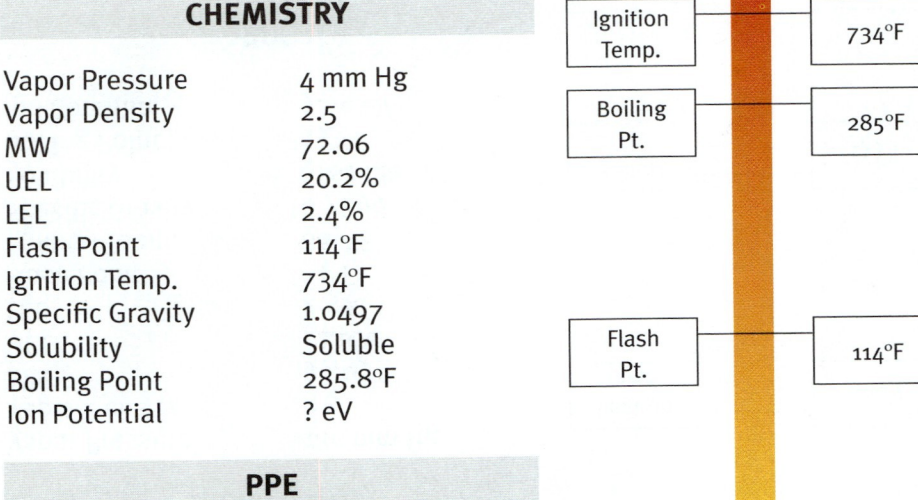

PPE

Depending on the quantity of the spilled material, wear the appropriate level of protection to prevent the possibility of skin contact and respiratory tract involvement. Avoid all direct physical contact by wearing protective ensemble with self-contained breathing apparatus.

>2 to <2,000 ppm	SAR
2,000 ppm and greater	SCBA

DECONTAMINATION

Wash IMMEDIATELY with soap and water, removing contaminated clothing. Constant irrigation to ensure removal of the contaminate from soft tissue and mucous membranes.

AMMONIA
NH₃

UN # 1005 Anhydrous
     2672 (10–35%)
     2073 (>35–50%)
     1005 (>50%)
CAS # 7664-41-7
RTECS # B00875000

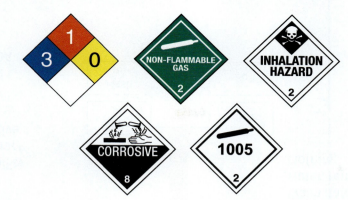

GUIDE # 125 (anhydrous)
      54 (10–35%)
     125 (>35–50%)
     125 (>50%)

**Poison Gas
Nonflammable
Corrosive
Inhalation Hazard**

DOT-ERG Suggestions:
Fire Isolation: 1,600 meters in all directions
Spills: Small: 100–200 meters isolation
        0.1 mile day/night downwind
       Large: 300 meters isolation
        0.3–0.7 miles day/night downwind

CHEMISTRY

Vapor Pressure	8.5 ATM
Vapor Density	.6
MW	17.03
UEL	28%
LEL	15%
Flash Point	52°F
Ignition Temp.	1204°F
Specific Gravity	
Solubility	47% Soluble
Boiling Point	−28°F
Ion Potential	10.18 eV

PPE

Depending on the quantity of the spilled material, wear the appropriate level of protection to prevent the possibility of skin contact and respiratory tract involvement, by wearing protective ensemble with self-contained breathing apparatus.

>50 to <300 ppm   SAR
300 ppm and greater   SCBA

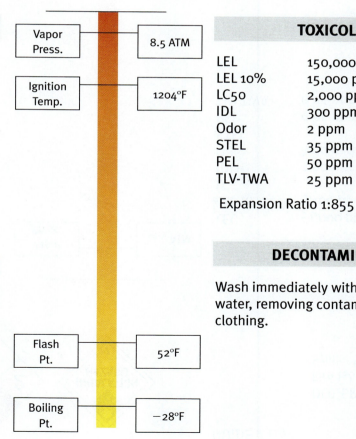

TOXICOLOGY

LEL	150,000 ppm	15%
LEL 10%	15,000 ppm	1.5%
LC50	2,000 ppm	0.2%
IDL	300 ppm	0.03%
Odor	2 ppm	0.0002%
STEL	35 ppm	0.0035%
PEL	50 ppm	0.005%
TLV-TWA	25 ppm	0.00025%

Expansion Ratio 1:855

DECONTAMINATION

Wash immediately with soap and water, removing contaminated clothing.

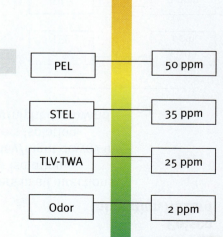

TOP 35 CHEMICALS/*Chemical Scan Sheets*

ARSINE
AsH$_3$

UN # 2188
CAS # 7784-42-1
RTECS # CG6475000
Military Designation: SA

GUIDE # 119

**CBRNE Chemical
Poison
Inhalation Hazard**

DOT-ERG Suggestions:
Fire Isolation: 800 Meters in all directions
Spills: Small: 60 meters Isolation
　　　　 0.3 –1.3 miles day/night downwind
　　　　 Large: 400 meters Isolation
　　　　 2.5–5 miles day/night downwind

CHEMISTRY

Vapor Pressure	14.9 ATM
Vapor Density	2.69
MW	77.93
UEL	78
LEL	5.1
Flash Point	−80°F
Ignition Temp.	
Specific Gravity	
Solubility	Slightly soluble
Boiling Point	−81°F
Ion Potential	9.89 eV

PPE

Wear the appropriate level of
protection to prevent the possibility
of skin contact and respiratory tract
involvement, by wearing protective
ensemble with self-contained
breathing apparatus.

>.05 to 3 ppm	SAR
3 ppm and greater	SCBA

TOXICOLOGY

LEL	51,000 ppm	5.1%
LEL 10%	5,100 ppm	.51%
LC$_{50}$		
IDLH	3 ppm	0.0003%
Odor	Mild garlic	
STEL		
PEL	0.05 ppm	0.000005%
TLV-TWA	0.25 ppm	0.000025%

DECONTAMINATION

Wash IMMEDIATELY with soap and
water, removing all contaminated
clothing.

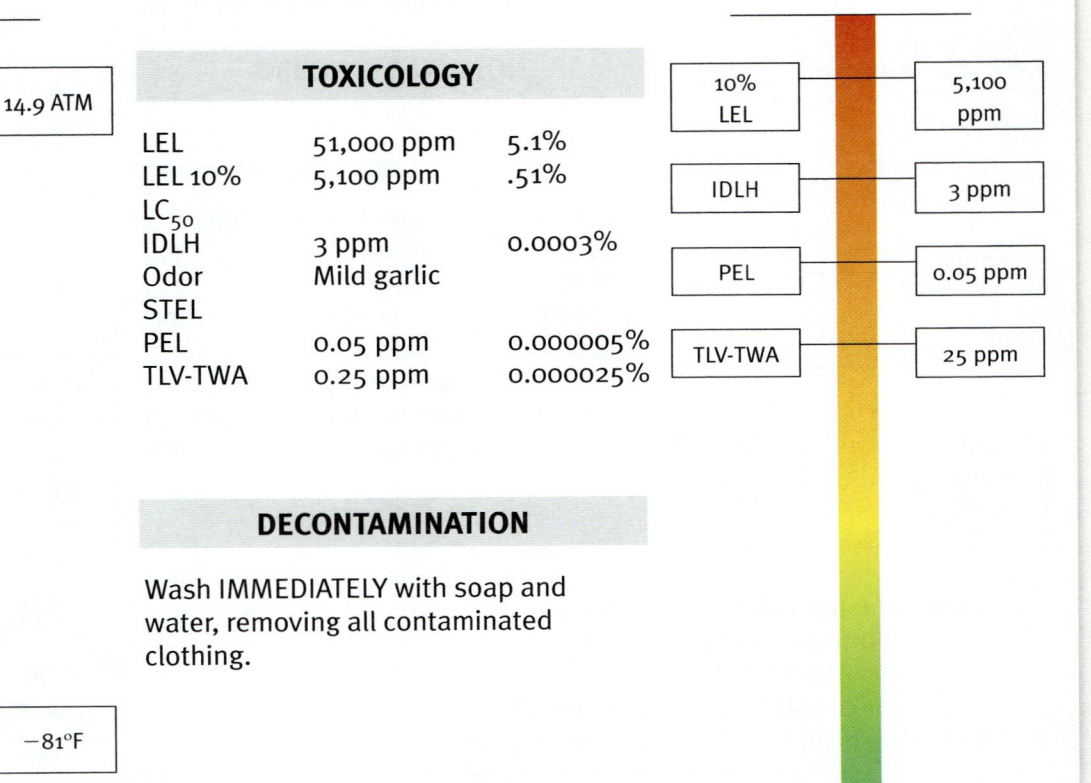

BENZENE
C$_6$H$_6$

UN # 1114
CAS # 71-43-2
RTECS # CY1400000

GUIDE # 130

Flammable

DOT-ERG Suggestions:
Fire Isolation: 800 meters in all directions
Spills: Small: 50–100 meters isolation
Large: 300 meters isolation

CHEMISTRY

Vapor Pressure	75 mm Hg
Vapor Density	2.7
MW	78.1
UEL	7.8%
LEL	1.2%
Flash Point	12°F
Ignition Temp.	928°F
Specific Gravity	0.88
Solubility	Insoluble
Boiling Point	176°F
Ion Potential	9.24 eV

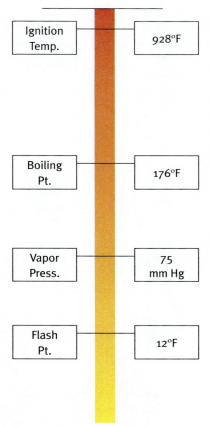

TOXICOLOGY

LEL	12,000 ppm	1.2%
LEL 10%	1,200 ppm	0.12%
LC50	10,000 ppm	1%
IDLH	500 ppm	0.05%
Odor	1 ppm	0.0001%
STEL	5 ppm	0.0005%
PEL	1 ppm	0.0001%
TLV-TWA	10 ppm	0.001%

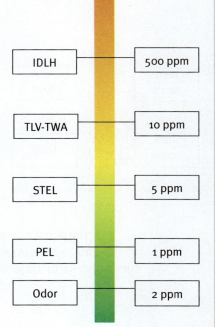

PPE

Depending on the quantity of the spilled material, wear the appropriate level of protection to prevent the possibility of skin contact and respiratory tract involvement, by wearing protective ensemble with self-contained breathing apparatus.

>1 to <100 ppm	APR
>100 to <1,000 ppm	SAR
1,000 ppm and greater	SCBA

DECONTAMINATION

Wash immediately with soap and water, removing contaminated clothing.

TOP 35 CHEMICALS / *Chemical Scan Sheets*

BUTADIENE
C_4H_6

UN # 1010 Inhibited
Polymerization
CAS # 106-99-0
RTECS # EI9275000

GUIDE # 116

**Polymerization
Flammable**

DOT-ERG Suggestions:
Fire Isolation: 1,600 meters in all directions
Spills: Small: 100+ meters isolation
Large: 800 meters isolation

CHEMISTRY

Vapor Pressure	2.42 ATM
Vapor Density	1.87
MW	54.09
UEL	11.5%
LEL	2.0%
Flash Point	−76°F
Ignition Temp.	788°F
Specific Gravity	0.62
Solubility	Insoluble
Boiling Point	24°F
Ion Potential	9.07 eV

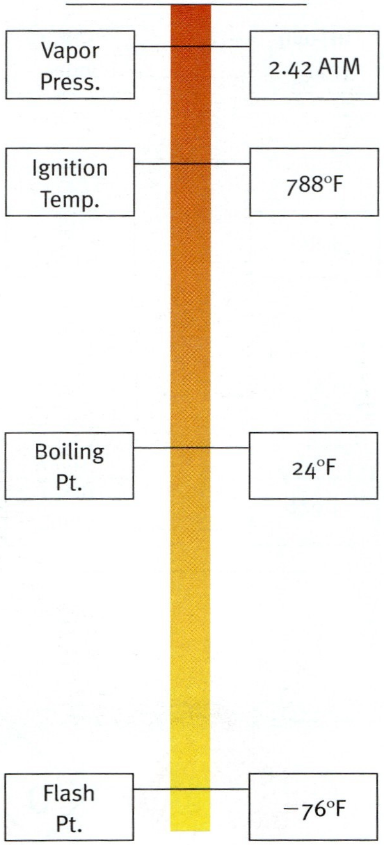

PPE

Eliminate all ignition sources. Depending on the quantity of the spilled material, wear the appropriate level of protection to prevent the possibility of skin contact and respiratory tract involvement, by wearing protective ensemble with self-contained breathing apparatus.

>1,000 to <2,000	SAR
2,000 ppm and greater	SCBA

Vapor Press.	2.42 ATM
Ignition Temp.	788°F
Boiling Pt.	24°F
Flash Pt.	−76°F

TOXICOLOGY

LEL	20,000 ppm	2.0%
LEL 10%	1,000 ppm	0.2%
LC_{50}	10,000 ppm	1%
IDLH	8,000 ppm	0.8%
Odor	1 ppm	0.0001%
STEL	5 ppm	0.0005%
PEL	1 ppm	0.0001%
TLV-TWA	10 ppm	0.001%

DECONTAMINATION

Wash immediately with soap and water, removing contaminated clothing.

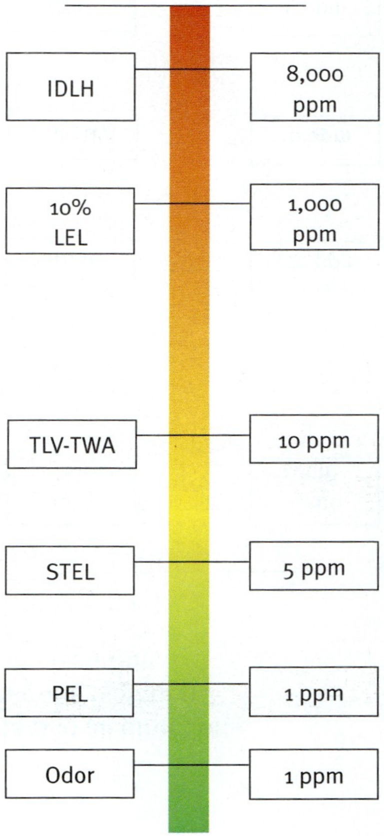

IDLH	8,000 ppm
10% LEL	1,000 ppm
TLV-TWA	10 ppm
STEL	5 ppm
PEL	1 ppm
Odor	1 ppm

3-QUINUCLIDINYL BENZILATE
$C_{21}H_{23}NO_3HCl$

UN # 2810
CAS # 6581-06-02
RTECS # VD6300000
Military Designation: BZ, (QNB)

GUIDE # 153

CBRNE Chemical
CNS Depressant

DOT-ERG Suggestions:
Fire Isolation: 800 meters in all directions
Spills: Small: 30 meters isolation
　　　　　　　0.1–0.2 miles day/night downwind
　　　　　Large: 200 meters isolation
　　　　　　　0.3–1.2 miles day/night downwind

CHEMISTRY

Vapor Pressure	Slight
Vapor Density	11
MW	337.41
UEL	
LEL	
Flash Point	474.8°F
Ignition Temp.	
Specific Gravity	
Solubility	Soluble
Boiling Point	608°F
Ion Potential	? eV

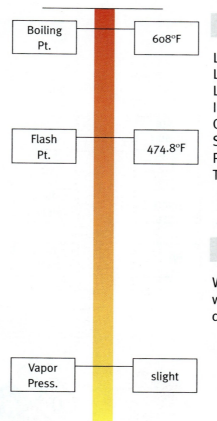

TOXICOLOGY

LEL	
LEL 10%	
LC50	200,000 mg-min/m³
IDLH	
Odor	None
STEL	
PEL	
TLV-TWA	~0.00026

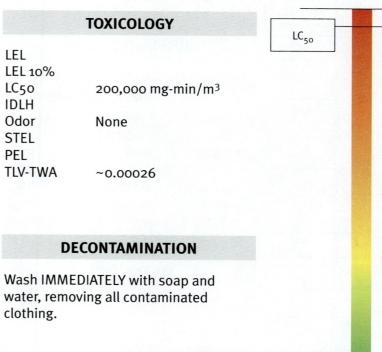

PPE

Wear the appropriate level of protection to prevent the possibility of skin contact and respiratory tract involvement, by wearing protective ensemble with self-contained breathing apparatus. Can cause unusual hallucination-type behavior.

DECONTAMINATION

Wash IMMEDIATELY with soap and water, removing all contaminated clothing.

CARBON MONOXIDE
CO

UN # 1016
CAS # 630-08-0
RTECS # FG3500000

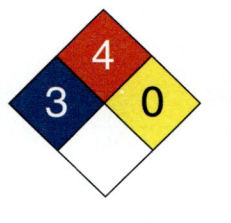

GUIDE # 119

Poison
Flammable

DOT-ERG Suggestions:
Fire Isolation: 1,600 meters in all directions
Spills: Small: 100–200 meters isolation
Large: 1,600 meters isolation

CHEMISTRY

Vapor Pressure	1 ATM
Vapor Density	0.968
MW	28.01
UEL	74%
LEL	12.5%
Flash Point	Flamm gas
Ignition Temp.	1128.2°F
Specific Gravity	
Solubility	insoluble
Boiling Point	−313°F
Ion Potential	14.01 eV

Vapor Press.	1 ATM
Ignition Temp.	1128°F
Flash Pt.	Flamm. Gas
Boiling Pt.	−313°F

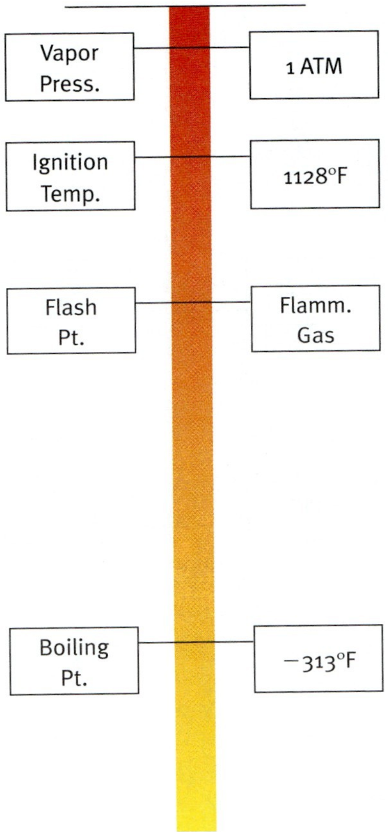

TOXICOLOGY

LEL	125,000 ppm	12.5%
LEL 10%	12,500 ppm	1.25%
LC50	5,000 ppm	0.5%
IDLH	1,200 ppm	0.12%
Odor		
STEL	400 ppm	0.04%
PEL	50 ppm	0.005%
TLV-TWA	25 ppm	0.0025%

Expansion Ratio 1:680

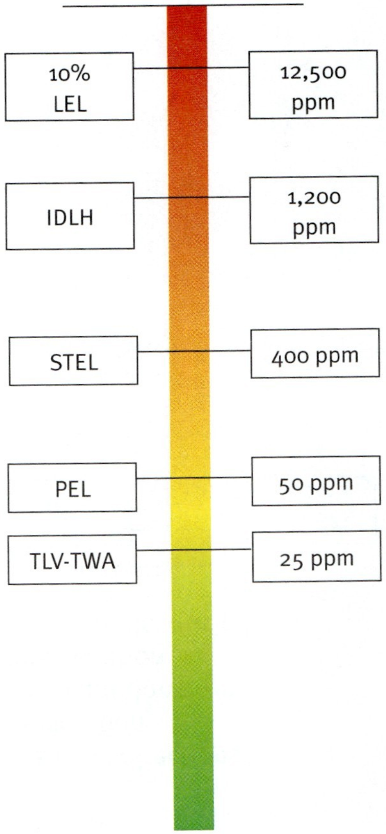

10% LEL	12,500 ppm
IDLH	1,200 ppm
STEL	400 ppm
PEL	50 ppm
TLV-TWA	25 ppm

DECONTAMINATION

Wash immediately with soap and water, removing contaminated clothing.

PPE

Eliminate all ignition sources. Depending on the quantity of the spilled material, wear the appropriate level of protection to respiratory tract involvement, by wearing protective ensemble with self-contained breathing apparatus.

>50 to <1,200 ppm	SAR
1,200 ppm and greater	SCBA

CYANOGEN CHLORIDE
ClC≡N

UN # 1589
CAS # 506-77-4
RTECS # GT2275000
Military Designation: CK

GUIDE # 125

CBRNE Chemical
Poison Gas
Inhalation Hazard

DOT-ERG Suggestions:
Fire Isolation: 1,600 meters in all directions
Spills: Small: 60 meters isolation
0.3–1.1 miles day/night downwind
Large: 275 meters isolation
1.7–4.2 miles day/night downwind

CHEMISTRY

Vapor Pressure	1010 mm Hg
Vapor Density	2.1
MW	61.48
UEL	
LEL	
Flash Point	Nonflammable
Ignition Temp.	
Specific Gravity	1.186
Solubility	Slightly
Boiling Point	55°F
Ion Potential	12.49 eV

PPE

Wear the appropriate level of protection to prevent the possibility of skin contact and respiratory tract involvement, by wearing protective ensemble with self-contained breathing apparatus.

>0.3 to 300 ppm	SAR
300 ppm and greater	SCBA

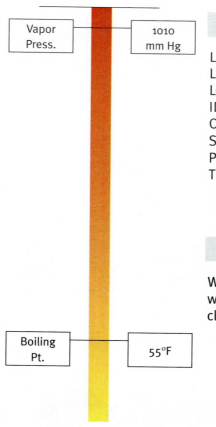

TOXICOLOGY

LEL		
LEL 10%		
LC$_{50}$		
IDLH		
Odor	0.71 ppm	0.000071%
STEL	0.3 ppm	0.00003%
PEL	0.3 ppm	0.00003%
TLV-TWA	0.3 ppm	0.00003%

DECONTAMINATION

Wash immediately with soap and water, removing contaminated clothing.

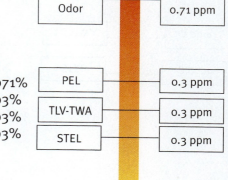

TOP 35 CHEMICALS / *Chemical Scan Sheets*

CHLORINE
Cl$_2$

UN # 1017
CAS # 7782-50-5
RTECS # FO2100000
Military Designation: CL

GUIDE # 124

CBRNE Chemical
Poison Gas
Corrosive
Inhalation Hazard

DOT-ERG Suggestions:
Fire Isolation: 800 meters in all directions
Spills: Small: 100–200 meters isolation
0.2–0.7 mile day/night downwind
Large: 300 meters isolation
1.7–4.2 miles day/night downwind

CHEMISTRY

Vapor Pressure	5 ATM
Vapor Density	2.5
MW	70.906
UEL	Supports combustion
LEL	
Flash Point	Nonflammable
Ignition Temp.	
Specific Gravity	1.424
Solubility	Reacts with water
Boiling Point	−30°F
Ion Potential	11.48 eV

Vapor Press. — 5 ATM

Boiling Pt. — −30°F

TOXICOLOGY

LEL		
LEL 10%		
LC$_{50}$	500 ppm	
0.05%		
IDLH	10 ppm	0.001%
Odor	0.31 ppm	0.000031%
STEL	3 ppm	0.0003%
PEL	0.5 ppm	0.00005%
TLV-TWA	1 ppm	0.0001%

Expansion Ratio 1:458

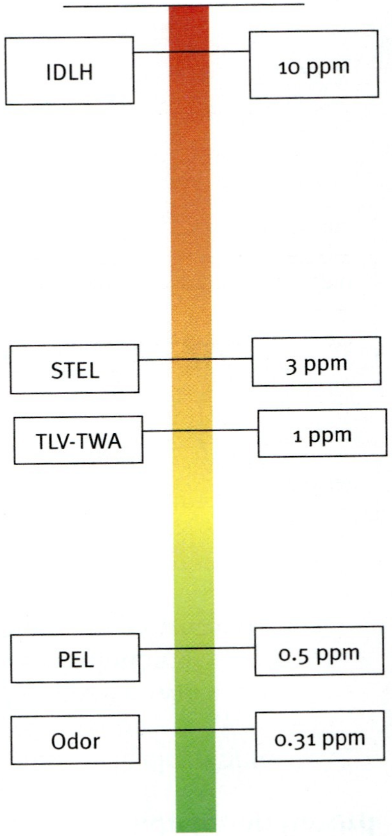

IDLH — 10 ppm

STEL — 3 ppm

TLV-TWA — 1 ppm

PEL — 0.5 ppm

Odor — 0.31 ppm

PPE

Depending on the quantity of the spilled material, wear the appropriate level of protection to prevent the possibility of skin contact and respiratory tract involvement, by wearing protective ensemble with self-contained breathing apparatus.

>1 to <10 ppm	SAR
10 ppm and greater	SCBA

DECONTAMINATION

Wash immediately with soap and water, removing contaminated clothing. Severe respiratory irritation with exposure.

ETHYL MERCAPTAN
C_2H_5SH

UN # 2363
CAS # 75-08-1
RTECS # KI9625000

GUIDE # 130

Flammable Gas

DOT-ERG Suggestions:
Fire Isolation: 800 meters in all directions
Spills: Small: 50–100 meters isolation
Large: 800 meters isolation

CHEMISTRY

Vapor Pressure	442 mm Hg
Vapor Density	2.14
MW	62.1
UEL	18%
LEL	2.8
Flash Point	−55°F
Ignition Temp.	570.2°F
Specific Gravity	0.826
Solubility	Slightly
Boiling Point	95°F
Ion Potential	9.29 eV

PPE

Depending on the quantity of the spilled material, wear the appropriate level of protection to prevent the possibility of skin contact and respiratory tract involvement, by wearing protective ensemble with self-contained breathing apparatus.

>10 to <100 ppm	SAR
100 ppm and greater	SCBA

TOXICOLOGY

LEL	28,000 ppm	2.8%
LEL 10%	2,800 ppm	0.28%
LC_{50}	4,420 ppm	0.44%
IDLH	500 ppm	0.05%
Odor	0.47 ppb	
STEL	10 ppm	0.001%
PEL	0.5 ppm	0.00005%
TLV-TWA	0.5 ppm	0.00005%

DECONTAMINATION

Wash immediately with soap and water, removing contaminated clothing.

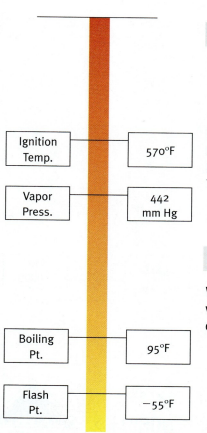

TOP 35 CHEMICALS / *Chemical Scan Sheets*

ETHYLDICHLOROARSINE
$C_2H_5AsCl_2$

UN # 1892
CAS # 598-14-1
RTECS # CH3500000
Military Designation: ED

GUIDE # 151

CBRNE Chemical
Poison
Inhalation Hazard
Corrosive

DOT-ERG Suggestions:
Fire Isolation: 800 meters in all directions
Spills: Small: 30 meters isolation
　　　　0.2 –5 miles day/night downwind
　　　Large: 125 meters isolation
　　　　0.8–1.6 miles day/night downwind

CHEMISTRY

Vapor Pressure	2.29 mm Hg
Vapor Density	6.0
MW	174.88
UEL	
LEL	
Flash Point	
Ignition Temp.	
Specific Gravity	1.66
Solubility	Soluble
Boiling Point	313°F
Ion Potential	? eV

PPE

Wear the appropriate level of protection to prevent the possibility of skin contact and respiratory tract involvement, by wearing protective ensemble with self-contained breathing apparatus. Do not touch unless wearing compatible protective equipment.

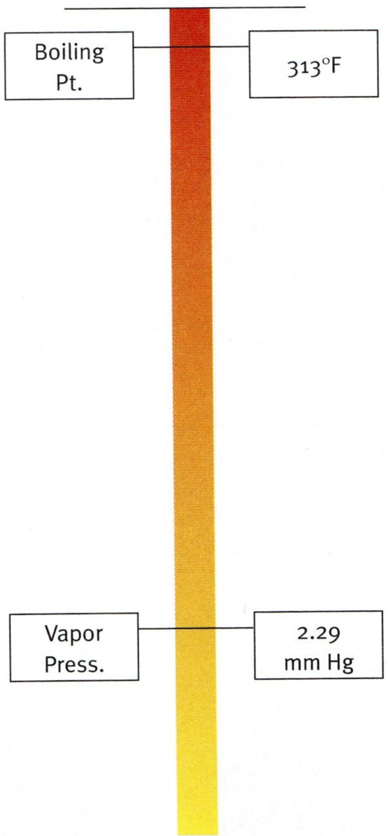

Boiling Pt. — 313°F

Vapor Press. — 2.29 mm Hg

TOXICOLOGY

LEL		
LEL 10%		
LC_{50}	14 ppm	0.00014%
IDLH		
Odor	Fairly fruity—biting	
STEL	0.002 ppm	0.0000002%
PEL	0.5 ppm	0.000005%
TLV-TWA	0.01 ppm	0.000001%

DECONTAMINATION

Wash IMMEDIATELY with soap and water, removing all contaminated clothing.

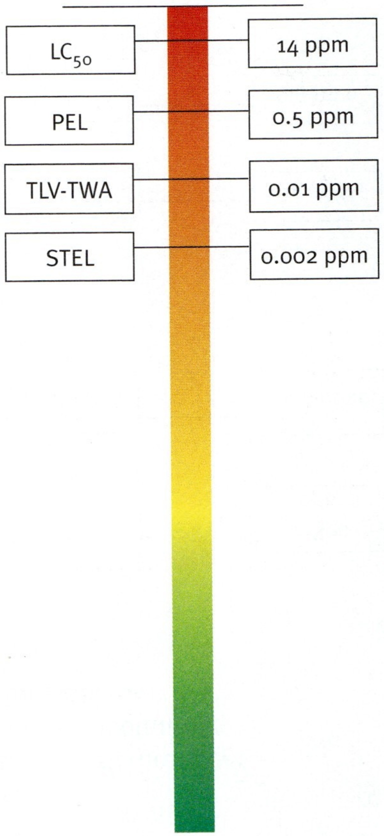

LC_{50} — 14 ppm
PEL — 0.5 ppm
TLV-TWA — 0.01 ppm
STEL — 0.002 ppm

ETHYLENE
$CH_2=CH_2$

UN # 1038 (cryogenic liq.)
　　　3138 (mixture)
　　　1962 (compressed)
CAS # 74-85-1
RTECS # KU5340000

GUIDE # 116P
115

Flammable Gas Polymerization

DOT-ERG Suggestions:
Fire Isolation: 1,600 meters in all directions
Spills: Small: 100+ meters isolation
　　　　Large: 800 meters isolation

CHEMISTRY

Vapor Pressure	47 ATM
Vapor Density	0.978
MW	28.05
UEL	36%
LEL	2.7%
Flash Point	−181°F
Ignition Temp.	914°F
Specific Gravity	
Solubility	Insoluble
Boiling Point	−152°F
Ion Potential	10.51 eV

PPE

Depending on the quantity of the spilled material, wear the appropriate level of protection to prevent the possibility of skin contact and respiratory tract involvement, by wearing protective ensemble with self-contained breathing apparatus.
All ranges　　　　　　　　SCBA

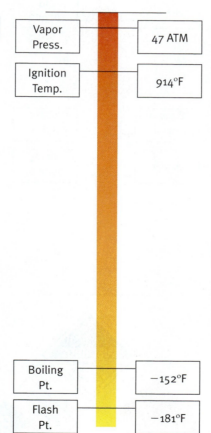

TOXICOLOGY

LEL	27,000 ppm	2.7%
LEL 10%	2,700 ppm	0.27%
LC_{50}		
IDLH		
Odor	260 ppm	0.026%
STEL		
PEL		
TLV-TWA	1,000 ppm	0.1%

DECONTAMINATION

Wash immediately with soap and water, removing contaminated clothing.

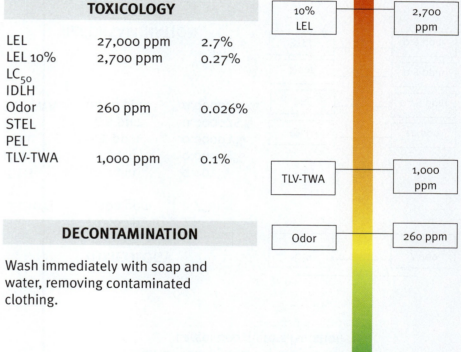

131

TOP 35 CHEMICALS/*Chemical Scan Sheets*

FORMALDEHYDE
HCHO

UN # 1198 (flammable)
   2209 (corrosive)
CAS # 50-00-0
RTECS # LP8925000

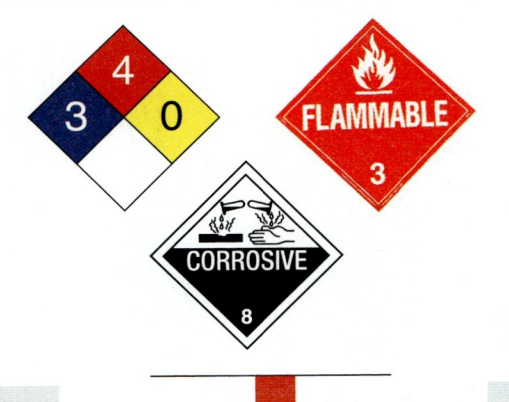

GUIDE # 132

Flammable Gas
Corrosive

DOT-ERG Suggestions:
Fire Isolation: 800 meters in all directions
Spills: Small: 50–100 meters isolation
   Large: 800 meters isolation

CHEMISTRY

Vapor Pressure	10–20 mm Hg
Vapor Density	1.075
MW	30.03
UEL	73%
LEL	7%
Flash Point	−55°F
Ignition Temp.	572°F
Specific Gravity	1.067
Solubility	Miscible
Boiling Point	−3°F
Ion Potential	10.88 eV

PPE

Depending on the quantity of the spilled material, wear the appropriate level of protection to prevent the possibility of skin contact and respiratory tract involvement, by wearing protective ensemble with self-contained breathing apparatus.

>0.75 to <20 ppm	SAR
20 ppm and greater	SCBA

Ignition Temp.	572°F
Vapor Press.	−20 mm Hg
Boiling Pt.	−3°F
Flash Pt.	−55°F

TOXICOLOGY

LEL	70,000 ppm	7%
LEL 10%	7,000 ppm	0.7%
LC$_{50}$		
IDLH	20 ppm	0.002%
Odor	0.5 ppm	0.00005%
STEL	0.3 ppm	0.00003%
PEL	0.75 ppm	0.000075%
TLV-TWA	0.016 ppm	0.0000016%

DECONTAMINATION

Wash immediately with soap and water, removing contaminated clothing.

10% LEL	7,000 ppm
IDLH	20 ppm
PEL	0.75 ppm
Odor	0.5 ppm
STEL	0.3 ppm
TLV-TWA	0.016 ppm

GASOLINE (Fuel Mixtures)

UN # 1203
CAS # 8006-61-9
RTECS # LX3300000

GUIDE # 128

Flammable

DOT-ERG Suggestions:
Fire Isolation: 800 meters in all directions
Spills: Small: 25–50 meters isolation
Large: 300 meters isolation

CHEMISTRY

Vapor Pressure	38–300 mm Hg
Vapor Density	3–4
MW	~72
UEL	7.6%
LEL	1.4%
Flash Point	−45°F
Ignition Temp.	852.8°F
Specific Gravity	0.7–0.8
Solubility	Insoluble
Boiling Point	77–428°F
Ion Potential	? eV

PPE

Depending on the quantity of the spilled material, wear the appropriate level of protection to prevent the possibility of skin contact and respiratory tract involvement, by wearing protective ensemble with self-contained breathing apparatus.

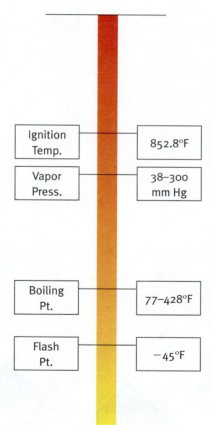

TOXICOLOGY

LEL	14,000 ppm	1.4%
LEL 10%	1,400 ppm	0.14%
LC_{50}		
IDLH		
Odor		
STEL	500 ppm	0.05%
PEL		
TLV-TWA	300 ppm	0.03%

DECONTAMINATION

Wash immediately with soap and water, removing contaminated clothing.

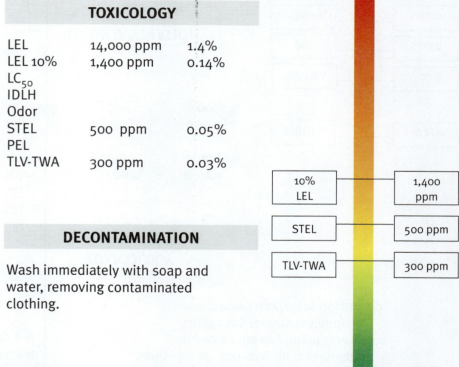

TOP 35 CHEMICALS/*Chemical Scan Sheets*

HYDROGEN CHLORIDE
HCl

UN # 1050 (anhydrous)
1789 (solution)
CAS # 7647-01-0
RTECS # MW4025000

GUIDE # 125 (anhydrous)
157 (solution)

Poison Gas
Corrosive
Inhalation Hazard

DOT-ERG Suggestions:
Fire Isolation: 1,600 meters in all directions
Spills: Small: 100–200 meters isolation
0.1–0.4 mile day/night downwind
Large: 185 meters isolation
1–2.7 miles day/night downwind

CHEMISTRY

Vapor Pressure	40.5 ATM
Vapor Density	1.268
MW	36.46
UEL	
LEL	
Flash Point	51.8°F
Ignition Temp.	
Specific Gravity	
Solubility	Miscible
Boiling Point	−121°F
Ion Potential	12.74 eV

PPE

Depending on the quantity of the spilled material, wear the appropriate level of protection to prevent the possibility of skin contact and respiratory tract involvement, by wearing protective ensemble with self-contained breathing apparatus.

>5 to <50 ppm	SAR
50 ppm and greater	SCBA

Vapor Press. — 40.5 ATM

Flash Pt. — 52°F

Boiling Pt. — −121°F

TOXICOLOGY

LEL		
LEL 10%		
LC_{50}		
IDLH	100 ppm	0.01%
Odor	5 ppm	0.0005%
STEL		
PEL	5 ppm	0.0005%
TLV-TWA	5 ppm	0.0005%

DECONTAMINATION

Wash immediately with soap and water, removing contaminated clothing. Severe burns to the mucous membranes and respiratory tract.

IDLH — 100 ppm
TLV-TWA — 5 ppm
PEL — 5 ppm
Odor — 5 ppm

HYDROGEN CYANIDE
HCN

UN # 1051 (>20%)
     1051 (anhydrous)
     1613 (<20%)
CAS # 74-90-8
RTECS # MW6825000
Military Designation: AC

GUIDE # 154
        117

**CBRNE Chemical
Poison Gas
Inhalation Hazard**

DOT-ERG Suggestions:
Fire Isolation: 800 meters in all directions
Spills: Small: 100 meters isolation
        0.1–0.3 mile day/night downwind
Large: 400 meters isolation
        0.8–2.1 miles day/night downwind

CHEMISTRY

Vapor Pressure	760 mm Hg
Vapor Density	0.901
MW	27.03
UEL	40%
LEL	5.6%
Flash Point	0°F
Ignition Temp.	1,000°F
Specific Gravity	0.68
Solubility	Miscible
Boiling Point	78.08°F
Ion Potential	13.60 eV

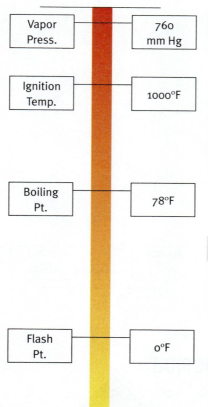

TOXICOLOGY

LEL	56,000 ppm	5.6%
LEL 10%	5,600 ppm	0.56%
LC_{50}	40 ppm	0.004%
IDLH	50 ppm	0.005%
Odor	0.81 ppm	0.000081%
STEL	4.7 ppm	0.00047%
PEL	10 ppm	0.001%
TLV-TWA	10 ppm	0.001%

DECONTAMINATION

Wash immediately with soap and water, removing contaminated clothing.

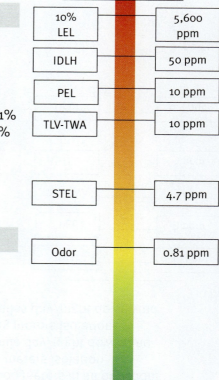

PPE

Wear the appropriate level of protection to prevent the possibility of skin contact and respiratory tract involvement, by wearing protective ensemble with self-contained breathing apparatus.

>10 to 50 ppm	SAR
50 ppm and greater	SCBA

HYDROGEN FLUORIDE
HF

UN #   1052 (anhydrous)
        1790 (solution)
CAS # 7664-39-3
RTECS # MW7875000

GUIDE # 116

Corrosive Poison

DOT-ERG Suggestions:
Fire Isolation: 1,600 meters in all directions
Spills: Small: 30 meters isolation
            0.1–0.4 mile day/night downwind
        Large: 125 meters isolation
            0.7–1.8 miles day/night downwind

CHEMISTRY

Vapor Pressure	760 mm Hg
Vapor Density	0.7
MW	20
UEL	
LEL	
Flash Point	
Ignition Temp.	
Specific Gravity	0.991
Solubility	Miscible
Boiling Point	67°F
Ion Potential	15.98 eV

Vapor Press. — 760 mm Hg

Boiling Pt. — 67°F

TOXICOLOGY

LEL		
LEL 10%		
LC$_{50}$	50 ppm	
0.005%		
IDLH	30 ppm	0.003%
Odor	0.5 ppm	0.00005%
STEL	3 ppm	0.0003%
PEL	3 ppm	0.0003%
TLV-TWA	3 ppm	0.0003%

IDLH — 30 ppm
PEL — 3 ppm
TLV-TWA — 3 ppm
STEL — 3 ppm
Odor — 0.5 ppm

DECONTAMINATION

Wash immediately with soap and
water, removing contaminated
clothing.

PPE

Wear the appropriate level of protection to
prevent the possibility of skin contact and
respiratory tract involvement, by wearing
protective ensemble with self-contained
breathing apparatus.

>3 to <30 ppm	SAR
30 ppm and greater	SCBA

HYDROGEN PEROXIDE
H_2O_2

UN # 2984 (8–20% solution)
    2014 (20–60% solution)
    2015 (>60% solution)
CAS # 7722-84-1
RTECS # MX0900000

GUIDE # 140
         143

Oxidizer
Flammable
Inhalation Hazard

DOT-ERG Suggestions:
Fire Isolation: 800 meters in all directions
Spills: Small: 10–25 meters isolation
        Large: 800 meters isolation

CHEMISTRY

Vapor Pressure	23.0 mm Hg
Vapor Density	1.02
MW	34.01
UEL	100%
LEL	40%
Flash Point	
Ignition Temp.	
Specific Gravity	1.39
Solubility	Miscible
Boiling Point	226.4°F
Ion Potential	10.54 eV

PPE

Wear the appropriate level of protection to prevent the possibility of skin contact and respiratory tract involvement, by wearing protective ensemble with self-contained breathing apparatus.

>1 to <75 ppm	SAR
75 ppm and greater	SCBA

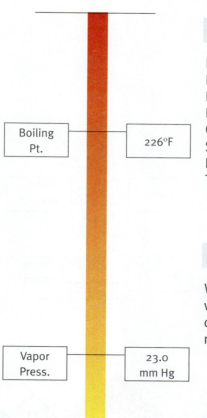

TOXICOLOGY

LEL	400,000 ppm	40%
LEL 10%	40,000 ppm	4%
LC_{50}		
IDLH	75 ppm	0.0075%
Odor		
STEL		
PEL	1 ppm	0.0001%
TLV-TWA	1 ppm	0.0001%

DECONTAMINATION

Wash immediately with soap and water, removing contaminated clothing. Corrosive to the mucous membranes. Very strong oxidizer.

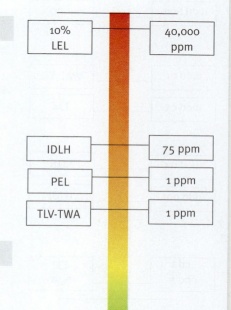

HYDROGEN SULFIDE
H₂S

UN # 1053
CAS # 7783-06-4
RTECS # MX1225000

GUIDE # 117

Poison Gas
Flammable
Inhalation Hazard

DOT-ERG Suggestions:
Fire Isolation: 1,600 meters in all directions
Spills: Small: 100–200 meters isolation
  0.1–0.2 mile day/night downwind
  Large: 215 meters isolation
  0.9–2.7 miles day/night downwind

CHEMISTRY

Vapor Pressure	17.6 ATM
Vapor Density	1.19
MW	34.1
UEL	44%
LEL	4%
Flash Point	
Ignition Temp.	500°F
Specific Gravity	0.916
Solubility	Slightly soluble
Boiling Point	−140.6°F
Ion Potential	10.46 eV

PPE

Depending on the quantity of the spilled material, wear the appropriate level of protection to prevent the possibility of skin contact and respiratory tract involvement, by wearing protective ensemble with self-contained breathing apparatus.

>20 to <100 ppm	SAR
100 ppm and greater	SCBA

Vapor Press. — 17.5 ATM

Ignition Temp. — 500°F

Boiling Pt. — −140°F

TOXICOLOGY

LEL	40,000 ppm	4%
LEL 10%	4,000 ppm	0.4%
LC₅₀	444 ppm	0.0444%
IDLH	100 ppm	0.01%
STEL	20 ppm	0.002%
PEL	10 ppm	0.001%
TLV-TWA	10 ppm	0.001%
Odor	0.05 ppm	0.00005%

DECONTAMINATION

Wash immediately with soap and water, removing contaminated clothing.

10% LEL — 4,000 ppm

IDLH — 100 ppm

STEL — 20 ppm

PEL — 10 ppm

TLV-TWA — 10 ppm

Odor — 0.05 ppm

LEWISITE
$C_2H_2AsCl_3$

UN # 2810
CAS # 541-25-3
RTECS # CH2975000
Military Designation: L

GUIDE # 153

**CBRNE Chemical
Poison
Inhalation Hazard**

DOT-ERG Suggestions:
Fire Isolation: 800 meters in all directions
Spills: Small: 30 meters isolation
      0.1–0.2 mile day/night downwind
   Large: 95 meters isolation
      0.6–1.1 miles day/night downwind

CHEMISTRY

Vapor Pressure	0.087 mm Hg
Vapor Density	7.1
MW	207.35
UEL	
LEL	
Flash Point	
Ignition Temp.	
Specific Gravity	1.89
Solubility	Insoluble
Boiling Point	374°F
Ion Potential	? eV

TOXICOLOGY

LEL	
LEL 10%	
LC_{50}	1,000 mg-min/m³
IDLH	
Odor	Geraniums
STEL	
PEL	0.003 mg-min/m³
TLV-TWA	0.003 mg-min/m³

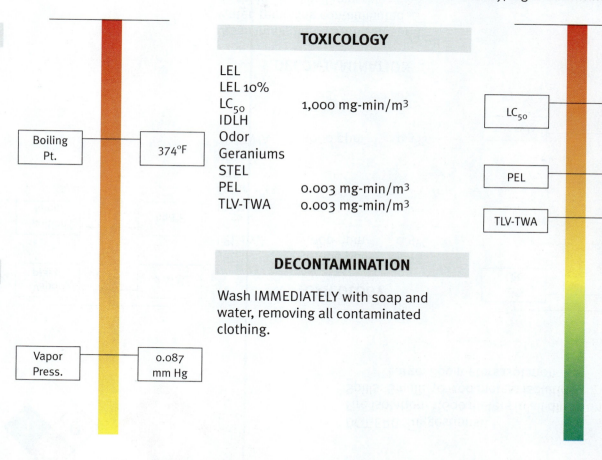

Boiling Pt. — 374°F
Vapor Press. — 0.087 mm Hg

LC_{50} — ~118 ppm
PEL — ~0.0005 ppm
TLV-TWA — ~0.0005 ppm

PPE

Wear the appropriate level of protection to prevent the possibility of skin contact and respiratory tract involvement, by wearing protective ensemble with self-contained breathing apparatus.

DECONTAMINATION

Wash IMMEDIATELY with soap and water, removing all contaminated clothing.

METHANE (Natural Gas)
CH_4

UN # 1203
CAS # 8006-61-9
RTECS # LX3300000

GUIDE # 115

Flammable

DOT-ERG Suggestions:
Fire Isolation: 1,600 meters in all directions
Spills: Small: 50–100 meters isolation
Large: 800 meters isolation

CHEMISTRY

Vapor Pressure	2 ATM
Vapor Density	0.55
MW	16.04
UEL	15%
LEL	5%
Flash Point	−306°F
Ignition Temp.	998.6°F
Specific Gravity	0.916
Solubility	Slightly soluble
Boiling Point	−259°F
Ion Potential	12.61 eV

PPE

Depending on the quantity of the spilled material, wear the appropriate level of protection to prevent the possibility of skin contact and respiratory tract involvement, by wearing protective ensemble with self-contained breathing apparatus.

Vapor Press. — 2 ATM

Ignition Temp. — 998°F

Boiling Pt. — −259°F

TOXICOLOGY

LEL	50,000 ppm	5%
LEL 10%	5,000 ppm	0.5%
LC_{50}		
IDLH		
Odor		
STEL		
PEL		
TLV-TWA	1,000 ppm	0.1%

DECONTAMINATION

Wash immediately with soap and water, removing contaminated clothing. Highly flammable.

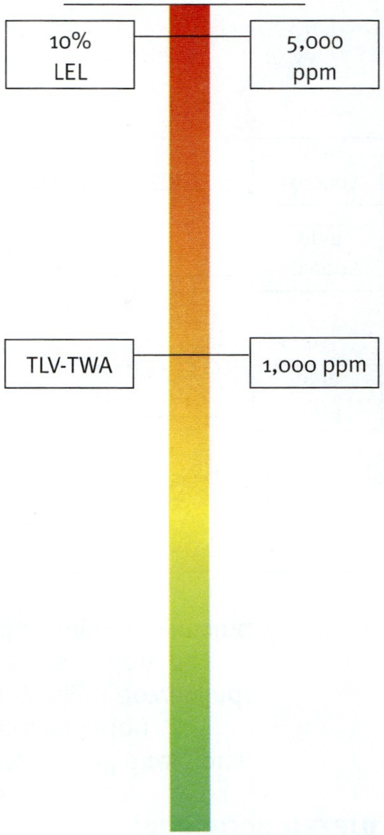

10% LEL — 5,000 ppm

TLV-TWA — 1,000 ppm

NAPHTHA (Coal Tar)

UN # 1256 (solvent)
    1255
    2553
    1268
CAS # 8030-30-6
RTECS # DE3030000

NFPA: Health 1, Flammability 4, Reactivity 0

GUIDE # 128

Flammable

DOT-ERG Suggestions:
Fire Isolation: 800 meters in all directions
Spills: Small: 25–50 meters isolation
       Large: 300 meters isolation

CHEMISTRY

Vapor Pressure	<5 mm Hg
Vapor Density	2.8
MW	Mixtures
UEL	6%
LEL	1%
Flash Point	107°F
Ignition Temp.	531°F
Specific Gravity	0.9
Solubility	Insoluble
Boiling Point	300°F
Ion Potential	? eV

PPE

Depending on the quantity of the spilled material, wear the appropriate level of protection to prevent the possibility of skin contact and respiratory tract involvement, by wearing protective ensemble with self-contained breathing apparatus.

>100 to <1,000 ppm	SAR
1,000 ppm and greater	SCBA

Ignition Temp. — 531°F
Boiling Pt. — 300°F
Flash Pt. — 107°F
Vapor Press. — <5 mm Hg

TOXICOLOGY

LEL	10,000 ppm	1%
LEL 10%	1,000 ppm	0.1%
LC$_{50}$		
IDLH	1,000 ppm	0.1%
Odor	100 ppm	0.01%
STEL		
PEL		
TLV-TWA	100 ppm	0.01%

DECONTAMINATION

Wash immediately with soap and water, removing contaminated clothing. Highly flammable; can cause mucous membranes damage.

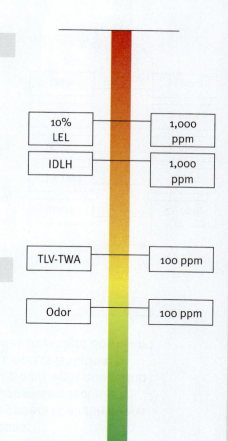

10% LEL — 1,000 ppm
IDLH — 1,000 ppm
TLV-TWA — 100 ppm
Odor — 100 ppm

TOP 35 CHEMICALS / *Chemical Scan Sheets*

NITRIC ACID
HNO_3

UN # 1760 (<40% acid)
     2031 (>40% acid)
     2032 (fuming)
CAS # 7697-37-2
RTECS # QU5775000

GUIDE # 154
        157

**Corrosive
Oxidizer**

DOT-ERG Suggestions:
Fire Isolation: 800 meters in all directions
Spills: Small: 25–100 meters isolation
                0.2–0.3 mile day/night downwind
       Large: 400 meters isolation
                0.8–2.2 miles day/night downwind

CHEMISTRY

Vapor Pressure	48 mm Hg
Vapor Density	2
MW	63.01
UEL	
LEL	
Flash Point	
Ignition Temp.	
Specific Gravity	1.55
Solubility	Soluble
Boiling Point	181°F
Ion Potential	11.95 eV

PPE

Wear the appropriate level of protection to prevent the possibility of skin contact and respiratory tract involvement, by wearing protective ensemble with self-contained breathing apparatus.

>2 to <25 ppm	SAR
25 ppm and greater	SCBA

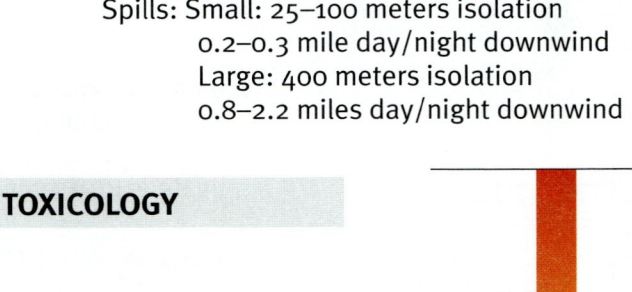

Boiling Pt. — 181°F

Vapor Press. — 48 mm Hg

TOXICOLOGY

LEL		
LEL 10%		
LC_{50}		
IDLH	25 ppm	0.0025%
Odor	0.31 ppm	0.000031%
STEL	4 ppm	0.0004%
PEL	2 ppm	0.0002%
TLV-TWA	2 ppm	0.0002%

DECONTAMINATION

Wash IMMEDIATELY with soap and water, removing contaminated clothing. Heavy exposures can cause severe lung damage. Corrosive to the mucous membranes.

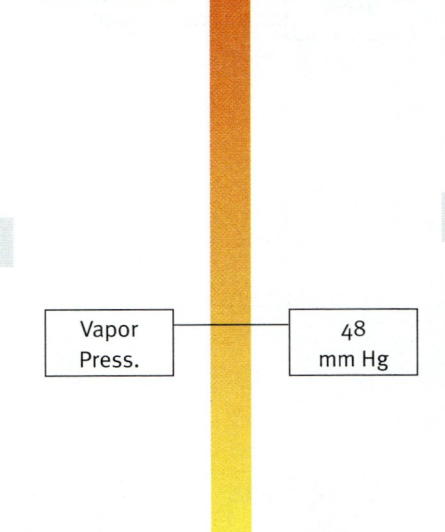

IDLH	25 ppm
TLV-TWA	2 ppm
STEL	4 ppm
PEL	2 ppm
Odor	31 ppm

NITROGEN MUSTARD
$C_6H_{13}Cl_2N$ (HN-1)

UN # 2810
CAS # 538-07-8 (HN-1)
        51-75-2 (HN-2)
        555-77-1 (HN-3)
RTECS # YE1225000 (HN-1)
         IA1750000 (HN-2)
         YE2625000 (HN-3)
Military Designation: HN

GUIDE # 153

**CBRNE Chemical
Poison
Inhalation Hazard**

DOT-ERG Suggestions:
Fire Isolation: 800 meters in all directions
Spills: Small: 30 meters isolation
           0.1 mile day/night downwind
        Large: 60 meters isolation
           0.4–0.8 mile day/night downwind

CHEMISTRY

Vapor Pressure	0.25–0.011 mm Hg
Vapor Density	5.9
MW	170.08
UEL	
LEL	
Flash Point	
Ignition Temp.	
Specific Gravity	1.09
Solubility	Slightly soluble
Boiling Point	381.2°F
Ion Potential	? eV

PPE

Wear the appropriate level of protection to prevent the possibility of skin contact and respiratory tract involvement, by wearing protective ensemble with self-contained breathing apparatus.

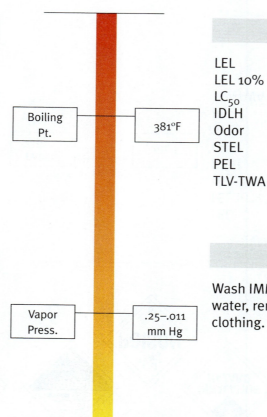

TOXICOLOGY

LEL	
LEL 10%	
LC_{50}	1,500 mg-min/m³
IDLH	
Odor	Fishy/musty
STEL	
PEL	0.003 mg-min/m³
TLV-TWA	0.003 mg-min/m³

DECONTAMINATION

Wash IMMEDIATELY with soap and water, removing all contaminated clothing.

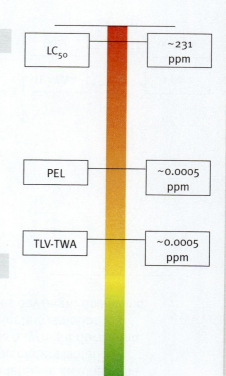

OLEUM
$SO_3-H_2SO_4$

UN # 1831
CAS # 8014-95-7
RTECS # WS5605000

4 1 2 W

CORROSIVE 8

POISON 6

INHALATION HAZARD 2

GUIDE # 137

**Poison
Corrosive**

DOT-ERG Suggestions:
Fire Isolation: 800 meters in all directions
Spills: Small: 50–100 meters isolation
0.2–0.7 mile day/night downwind
Large: 300 meters isolation
1.3–3.5 miles day/night downwind

CHEMISTRY

Vapor Pressure	2 mm Hg
Vapor Density	2.76
MW	178.14
UEL	
LEL	
Flash Point	
Ignition Temp.	
Specific Gravity	11.94
Solubility	Reacts in water
Boiling Point	284°F
Ion Potential	? eV

PPE

Wear the appropriate level of protection to prevent the possibility of skin contact and respiratory tract involvement, by wearing protective ensemble with self-contained breathing apparatus.

Boiling Pt. — 284°F

Vapor pressure — 2 mm Hg

TOXICOLOGY

LEL		
LEL 10%		
LC_{50}		
IDLH		
Odor	0.1 ppm	0.00001%
STEL		
PEL		
TLV-TWA	1 ppm	0.0001%

DECONTAMINATION

Wash IMMEDIATELY with soap and water, removing contaminated clothing. Corrosive to the mucous membranes and extremely toxic. May cause blindness. Extremely hydroscopic.

TLV-TWA — 1 ppm

Odor — 0.1 ppm

PHENOL
C_6H_5OH

UN # 1671 (solid)
   2312 (molten)
   2821 (solution)
CAS # 108-95-2
RTECS # SJ3325000

GUIDE # 153

Poison

DOT-ERG Suggestions:
Fire Isolation: 800 meters in all directions
Spills: Small: 25–50 meters isolation
       Large: 800 meters isolation

CHEMISTRY

Vapor Pressure	0.4 mm Hg
Vapor Density	3.24
MW	94.1
UEL	8.6%
LEL	1.7%
Flash Point	175°F
Ignition Temp.	1,320°F
Specific Gravity	1.05
Solubility	Slightly
Boiling Point	359°F
Ion Potential	8.50 eV

PPE

Depending on the quantity of the spilled material, wear the appropriate level of protection to prevent the possibility of skin contact and respiratory tract involvement, by wearing protective ensemble with self-contained breathing apparatus.

>5 to <250 ppm   SAR
250 ppm and greater   SCBA

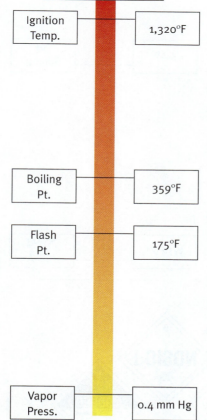

TOXICOLOGY

LEL	17,000 ppm	1.7%
LEL 10%	1,700 ppm	0.17%
LC_{50}		
IDLH	250 ppm	0.025%
Odor	0.048 ppm	0.000004%
STEL	15 ppm	0.0015%
PEL	5 ppm	0.0005%
TLV-TWA	5 ppm	0.0005%

DECONTAMINATION

Wash IMMEDIATELY with soap and water, removing contaminated clothing. Corrosive to the mucous membranes, and absorbed through the skin. Can cause severe neurological effects.

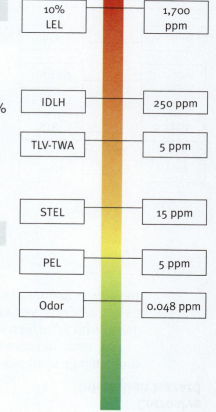

TOP 35 CHEMICALS / *Chemical Scan Sheets*

PHOSGENE
COCl$_2$

UN # 1076
CAS # 75-44-5
RTECS # SY5600000
Military Designation: CG

GUIDE # 125

**CBRNE Chemical
Poison Gas
Corrosive
Inhalation Hazard**

DOT-ERG Suggestions:
Fire Isolation: 1,600 meters in all directions
Spills: Small: 100–200 meters isolation
0.5–1.7 miles day/night downwind
Large: 765 meters isolation
4.2–6.9 miles day/night downwind

CHEMISTRY

Vapor Pressure	1.6 ATM
Vapor Density	3.4
MW	98.92
UEL	
LEL	
Flash Point	Nonflammable
Ignition Temp.	
Specific Gravity	1.38
Solubility	Slightly soluble
Boiling Point	47°F
Ion Potential	11.55 eV

PPE

Wear the appropriate level of protection to prevent the possibility of skin contact and respiratory tract involvement, by wearing protective ensemble with self-contained breathing apparatus.

>0.1 to 2 ppm	SAR
2 ppm and greater	SCBA

TOXICOLOGY

LEL		
LEL 10%		
LC$_{50}$	50 ppm	
0.005%		
IDLH	2 ppm	0.0002%
Odor	0.5 ppm	0.00005%
STEL	0.2 ppm	0.00002%
PEL	0.1 ppm	0.00001%
TLV-TWA	0.1 ppm	0.00001%

DECONTAMINATION

Wash immediately with soap and water, removing contaminated clothing. Can cause severe burns to the mucous membranes and extreme respiratory deficiencies.

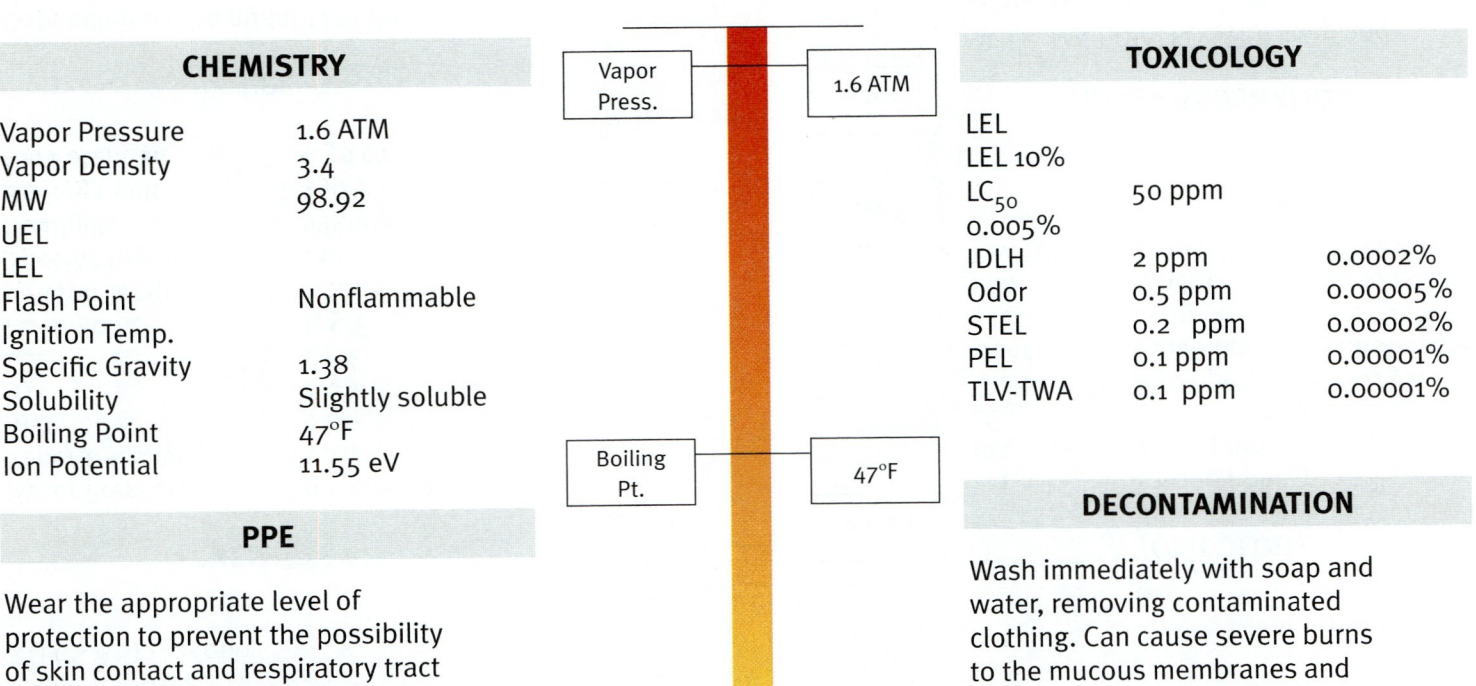

PHOSGENE OXIME
$CHCl_2NO$

UN # 2811
CAS # 1794-86-1
RTECS #
Military Designation: CX

GUIDE # 154

**CBRNE Chemical
Poison
Inhalation Hazard**

DOT-ERG Suggestions:
Fire Isolation: 800 meters in all directions
Spills: Small: 30 meters isolation
        0.1–0.2 mile day/night downwind
     Large: 95 meters isolation
        0.6–1.1 miles day/night downwind

CHEMISTRY

Vapor Pressure	11.2 mm Hg
Vapor Density	3.9
MW	113.9
UEL	
LEL	
Flash Point	
Ignition Temp.	
Specific Gravity	
Solubility	Slightly soluble
Boiling Point	264.2°F
Ion Potential	? eV

PPE

Wear the appropriate level of protection to prevent the possibility of skin contact and respiratory tract involvement, by wearing protective ensemble with self-contained breathing apparatus.

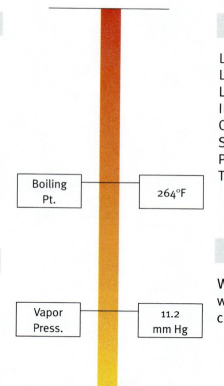

Boiling Pt. — 264°F

Vapor Press. — 11.2 mm Hg

TOXICOLOGY

LEL	
LEL 10%	
LC_{50}	1,500 mg-min/m³
IDLH	
Odor	Intense irritation
STEL	
PEL	
TLV-TWA	

LC_{50} — ~322 ppm

DECONTAMINATION

Wash IMMEDIATELY with soap and water, removing all contaminated clothing.

TOP 35 CHEMICALS / *Chemical Scan Sheets*

PHOSPHORIC ACID
H_3PO_4

UN # 1805
CAS # 7664-38-2
RTECS # TB6300000

3 0 0

CORROSIVE 8

GUIDE # 154

Corrosive

DOT-ERG Suggestions:
Fire Isolation: 800 meters in all directions
Spills: Small: 25–50 meters isolation
Large: 800 meters isolation

CHEMISTRY

Vapor Pressure	0.03 mm Hg
Vapor Density	3.4
MW	98
UEL	
LEL	
Flash Point	Nonflammable
Ignition Temp.	
Specific Gravity	1.874
Solubility	Miscible
Boiling Point	415°F
Ion Potential	? eV

PPE

Wear the appropriate level of protection to prevent the possibility of skin contact and respiratory tract involvement. Avoid all direct physical contact wearing protective ensemble with self-contained breathing apparatus.

>1 to 100 mg/m³	APR
>100 to <1,000 mg/m³	SAR
1,000 mg/m³ and greater	SCBA

TOXICOLOGY

LEL		
LEL 10%		
LC_{50}	~212 ppm	0.0212%
IDLH	~250 ppm	0.025%
Odor		
STEL	~0.75 ppm	0.000075%
PEL	~0.25 ppm	0.000025%
TLV-TWA	~0.25 ppm	0.000025%

DECONTAMINATION

Wash IMMEDIATELY with soap and water, removing contaminated clothing. Heavy exposures can cause severe lung damage. Corrosive to the mucous membranes.

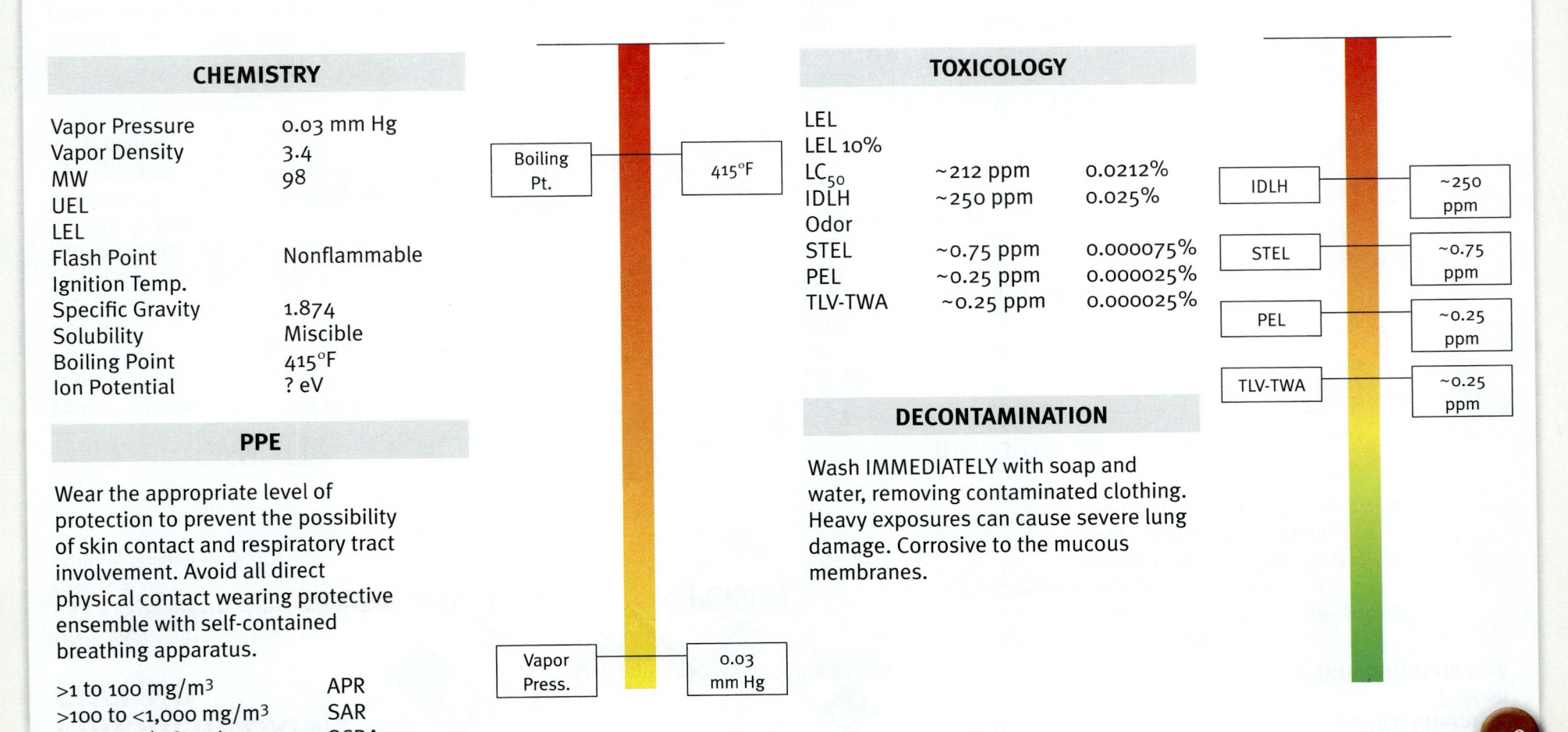

Boiling Pt. — 415°F

Vapor Press. — 0.03 mm Hg

IDLH — ~250 ppm

STEL — ~0.75 ppm

PEL — ~0.25 ppm

TLV-TWA — ~0.25 ppm

POTASSIUM HYDROXIDE
KOH

UN # 1813 (dry, solid)
      1814 (solution)
CAS # 1310-58-3
RTECS # TT2100000

GUIDE # 154

Corrosive

DOT-ERG Suggestions:
Fire Isolation: 800 meters in all directions
Spills: Small: 25–50 meters isolation
      Large: 800 meters isolation

CHEMISTRY

Vapor Pressure	1 mm Hg
Vapor Density	
MW	56.10
UEL	
LEL	
Flash Point	Nonflammable
Ignition Temp.	
Specific Gravity	2.044
Solubility	Soluble
Boiling Point	2415°F
Ion Potential	N/A

Boiling Pt. — 2415°F

TOXICOLOGY

LEL	
LEL 10%	
LC_{50}	
IDLH	
Odor	
STEL	2 mg/m³
PEL	2 mg/m³
TLV-TWA	2 mg/m³

PPE

Depending on the quantity of the spilled material, wear the appropriate level of protection to prevent the possibility of respiratory tract involvement, by wearing a protective ensemble.

>2 to 200 mg/m³	APR
>200 to 2,000 mg/m³	SAR
2,000 mg/m³ or greater	SCBA

DECONTAMINATION

Wash IMMEDIATELY with soap and water, removing contaminated clothing. Corrosive to the mucous membranes, and lung tissue.

Vapor Press. — 1 mm Hg

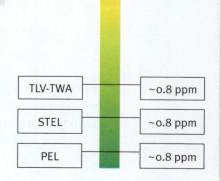

TLV-TWA — ~0.8 ppm
STEL — ~0.8 ppm
PEL — ~0.8 ppm

TOP 35 CHEMICALS / *Chemical Scan Sheets*

POTASSIUM PERMANGENATE
KMnO$_4$

UN # 1490
CAS # 7722-64-7
RTECS # SD6475000

GUIDE # 140

Oxidizer

DOT-ERG Suggestions:
Fire Isolation: 800 meters in all directions
Spills: Small: 25–50 meters isolation
Large: 800 meters isolation

CHEMISTRY

Vapor Pressure	
Vapor Density	
MW	158.04
UEL	
LEL	
Flash Point	Nonflammable
Ignition Temp.	
Specific Gravity	2.7
Solubility	Soluble in alcohol
Boiling Point	Decomposes 1 ATM
Ion Potential	N/A

Boiling Pt.

Decomposes at 1 ATM

PPE

Depending on the quantity of the spilled material, wear the appropriate level of protection to prevent the possibility of skin and respiratory tract involvement, by wearing a protective ensemble.

TOXICOLOGY

LEL	
LEL 10%	
LC$_{50}$	
IDLH	
Odor	
STEL	5 mg/m^3
PEL	5 mg/m^3
TLV-TWA	0.2 mg/m^3

DECONTAMINATION

Wash immediately with soap and water, removing contaminated clothing. Highly corrosive; causes burns to the mucous membranes.

STEL	~0.7 ppm
PEL	~0.7 ppm
TLV-TWA	~0.03 ppm

PROPANE
C_3H_8

UN # 1075
     1978
CAS # 74-98-6
RTECS # TX2275000

GUIDE # 115   **Flammable**

DOT-ERG Suggestions:
Fire Isolation: 1,600 meters in all directions
Spills: Small: 50–100 meters isolation
        Large: 1,600 meters isolation

CHEMISTRY

Vapor Pressure	8.4 ATM
Vapor Density	1.56
MW	44.1
UEL	9.5%
LEL	2.1%
Flash Point	−155°F
Ignition Temp.	842°F
Specific Gravity	0.5853
Solubility	Slight
Boiling Point	−44°F
Ion Potential	11.07 eV

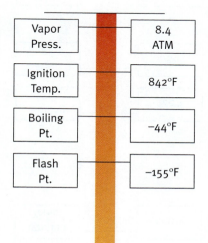

TOXICOLOGY

LEL	21,000 ppm	2.1%
LEL 10%	2,100 ppm	0.21%
LC_{50}		
IDLH	2,100 ppm	0.21%
Odor	5,000 ppm	0.5%
STEL		
PEL	1,000 ppm	0.1%
TLV-TWA	1,000 ppm	0.1%

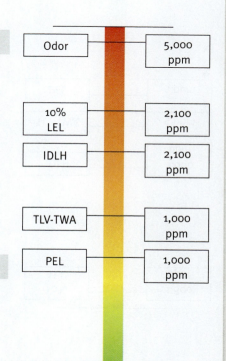

PPE

Wear the appropriate level of protection to prevent the possibility of respiratory tract involvement, by wearing protective ensemble with self-contained breathing apparatus.

>1,000 to <2,100 ppm	SAR
2,100 ppm and above	SCBA

DECONTAMINATION

Wash immediately with soap and water, removing contaminated clothing. Eliminate all ignition sources; can cause asphyxiation.

TOP 35 CHEMICALS / *Chemical Scan Sheets*

PYRIDINE
C_5H_5N

UN # 1282
CAS # 110-86-1
RTECS # UR8400000

GUIDE # 129

Flammable Poison

DOT-ERG Suggestions:
Fire Isolation: 800 meters in all directions
Spills: Small: 50–100 meters isolation
Large: 300 meters isolation

CHEMISTRY

Vapor Pressure	16 mm Hg
Vapor Density	2.72
MW	79.10
UEL	12.4%
LEL	1.8%
Flash Point	68°F
Ignition Temp.	900°F
Specific Gravity	0.98272
Solubility	Miscible
Boiling Point	240°F
Ion Potential	9.27 eV

PPE

Wear the appropriate level of protection to prevent the possibility of respiratory tract involvement, by wearing protective ensemble with self-contained breathing apparatus.

>5 to 500 ppm	APR
>500 to <1,000 ppm	SAR
1,000 ppm and greater	SCBA

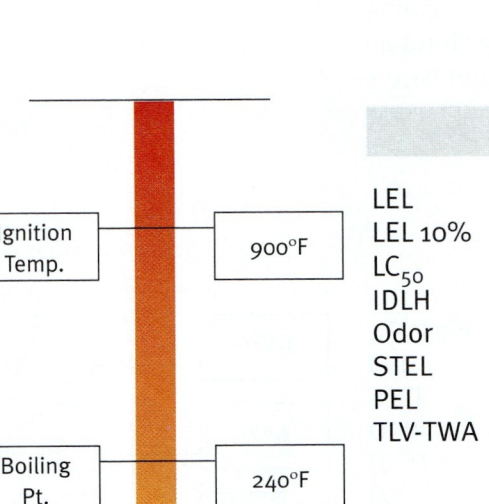

Ignition Temp.		900°F
Boiling Pt.		240°F
Flash Pt.		68°F
Vapor Press.		16 mm Hg

TOXICOLOGY

LEL	18,000 ppm	1.8%
LEL 10%	1,800 ppm	0.21%
LC_{50}		
IDLH	1,000 ppm	0.1%
Odor	21 ppb	
STEL		
PEL	5 ppm	0.0005%
TLV-TWA	5 ppm	0.0005%

DECONTAMINATION

Wash immediately with soap and water, removing contaminated clothing. Irritating to the mucous membranes and respiratory tract.

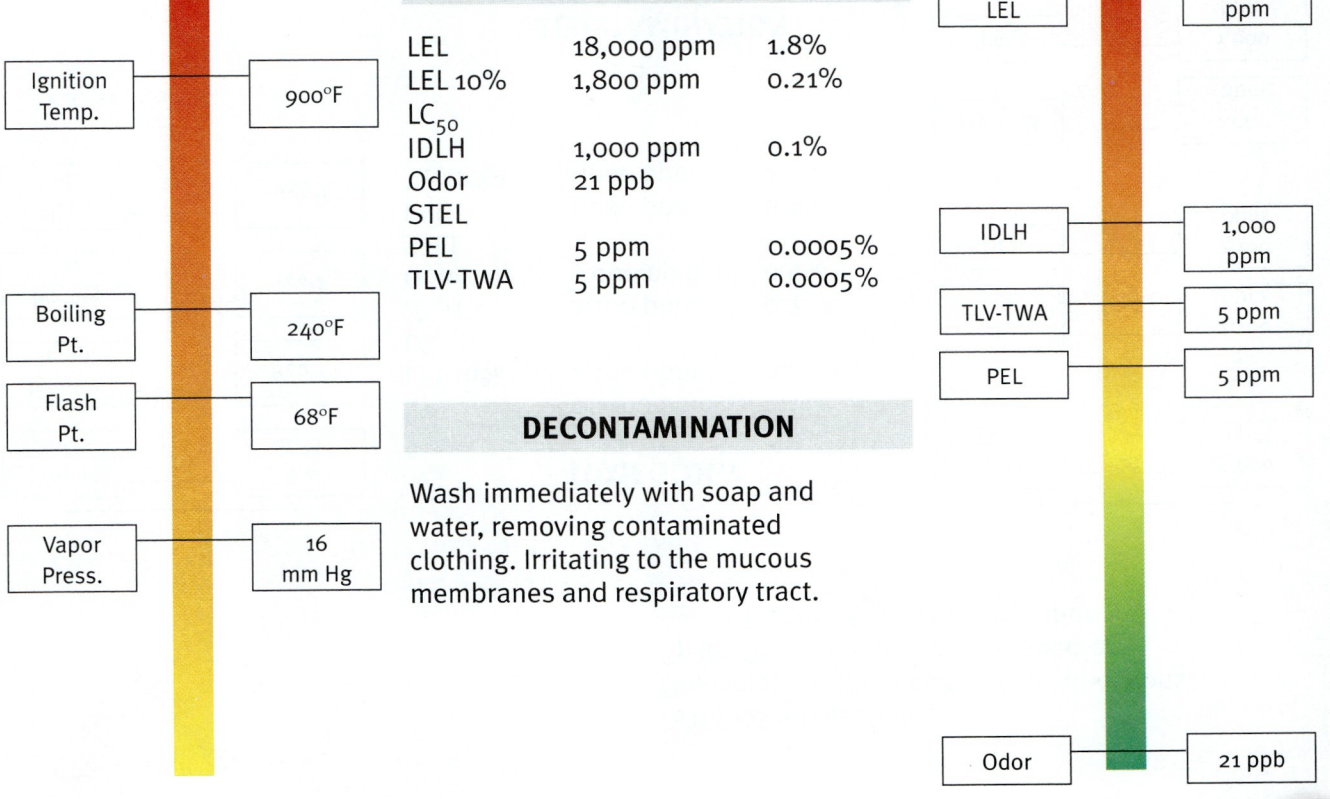

10% LEL		1,800 ppm
IDLH		1,000 ppm
TLV-TWA		5 ppm
PEL		5 ppm
Odor		21 ppb

SARIN
$C_4H_{10}FO_2P$

UN # 2810
CAS # 107-44-8
RTECS # TA8400000
Military Designation: GB

GUIDE # 153

CBRNE Chemical
Poison
Inhalation Hazard

DOT-ERG Suggestions:
Fire Isolation: 800 meters in all directions
Spills: Small: 155 meters isolation
 1–2.1 miles day/night downwind
 Large: 915 meters isolation
 7+ miles day/night downwind

CHEMISTRY

Vapor Pressure	2.1 mm Hg
Vapor Density	4.86
MW	140.09
UEL	
LEL	
Flash Point	Nonflammable
Ignition Temp.	
Specific Gravity	1.1
Solubility	Miscible
Boiling Point	297°F
Ion Potential	? eV

PPE

Wear the appropriate level of protection to prevent the possibility of skin contact and respiratory tract involvement, by wearing protective ensemble with self-contained breathing apparatus.

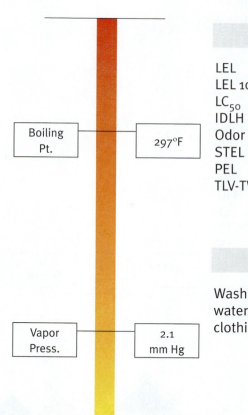

TOXICOLOGY

LEL	
LEL 10%	
LC_{50}	50–100 mg-min/m³
IDLH	
Odor	Faintly fruity
STEL	
PEL	0.0001 mg-min/m³
TLV-TWA	0.0001 mg-min/m³

DECONTAMINATION

Wash IMMEDIATELY with soap and water, removing all contaminated clothing.

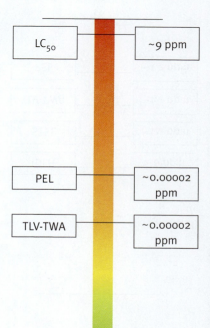

153

TOP 35 CHEMICALS/*Chemical Scan Sheets*

SODIUM HYDROXIDE
NaOH

UN # 1823 (solid)
        1824 (solution)
CAS # 1310-73-2
RTECS # WB4900000

GUIDE # 154

Corrosive

DOT-ERG Suggestions:
Fire Isolation: 800 meters in all directions
Spills: Small: 25–50 meters isolation
        Large: 800 meters isolation

CHEMISTRY

Vapor Pressure	0 mm Hg
Vapor Density	
MW	40.01
UEL	
LEL	
Flash Point	
Ignition Temp.	
Specific Gravity	2.13
Solubility	Miscible
Boiling Point	2534°F
Ion Potential	N/A

Boiling Pt. — 2534°F

PPE

Wear the appropriate level of protection to prevent the possibility of skin contact and respiratory tract involvement, by wearing protective ensemble with self-contained breathing apparatus.

>2 to <10 mg/m^3	APR
10 mg/m^3 and above	SCBA

TOXICOLOGY

LEL	
LEL 10%	
LC$_{50}$	
IDLH	10 mg/m^3
Odor	
STEL	2 mg/m^3
PEL	2 mg/m^3
TLV-TWA	2 mg/m^3

DECONTAMINATION

Wash immediately with soap and water, removing contaminated clothing. Due to the corrosive nature of the material, serious burns can occur.

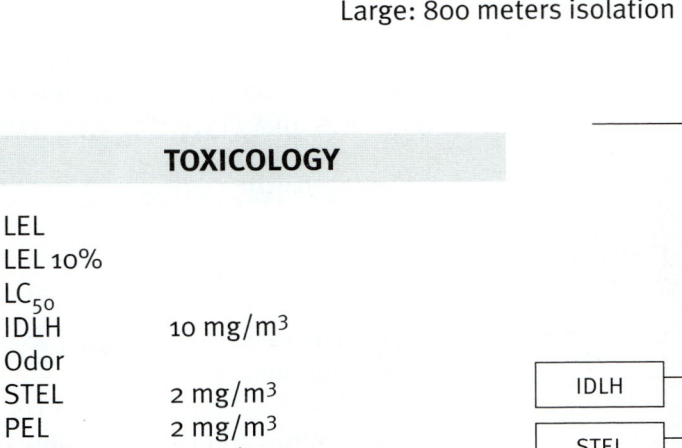

IDLH	~6 ppm
STEL	~1.2 ppm
TLV-TWA	~1.2 ppm
PEL	~1.2 ppm

154

SOMAN
$C_7H_{16}FO_2P$

UN # 2810
CAS # 96-64-0
RTECS # TA8750000
Military Designation: GD

GUIDE # 153

**CBRNE Chemical
Poison
Inhalation Hazard**

DOT-ERG Suggestions:
Fire Isolation: 800 meters in all directions
Spills: Small: 95 meters isolation
        0.5–1.1 miles day/night downwind
    Large: 765 meters isolation
        4.2–6.5 miles day/night downwind

CHEMISTRY

Vapor Pressure	0.4 mm Hg
Vapor Density	6.33
MW	182.2
UEL	
LEL	
Flash Point	
Ignition Temp.	
Specific Gravity	1.022
Solubility	Slightly soluble
Boiling Point	388.4°F
Ion Potential	? eV

PPE

Wear the appropriate level of protection to prevent the possibility of skin contact and respiratory tract involvement, by wearing protective ensemble with self-contained breathing apparatus.

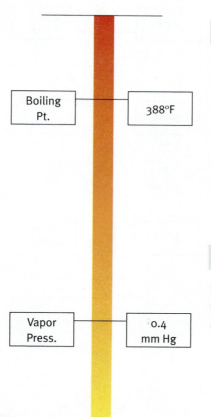

TOXICOLOGY

LEL	
LEL 10%	
LC_{50}	5–50 mg-min/m³
IDLH	
Odor	Fairly fruity
STEL	
PEL	0.00003 mg-min/m³
TLV-TWA	0.00003 mg-min/m³

DECONTAMINATION

Wash IMMEDIATELY with soap and water, removing all contaminated clothing.

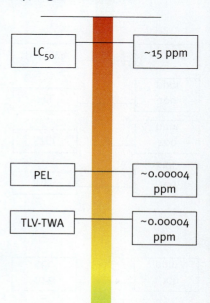

TOP 35 CHEMICALS/*Chemical Scan Sheets*

STYRENE
$C_6H_5CH=CH_2$

UN # 2055 (inhibited)
CAS # 100-42-5
RTECS # WL3675000

GUIDE # 128P

Polymerization Flammable

DOT-ERG Suggestions:
Fire Isolation: 800 meters in all directions
Spills: Small: 25–50 meters isolation
Large: 300 meters isolation

CHEMISTRY

Vapor Pressure	9.5 mm Hg
Vapor Density	1.1
MW	104.15
UEL	7%
LEL	1.1%
Flash Point	88°F
Ignition Temp.	914°F
Specific Gravity	0.906
Solubility	Sparingly soluble
Boiling Point	293°F
Ion Potential	8.4 eV

PPE

Depending on the quantity of the spilled material, wear the appropriate level of protection to prevent the possibility of skin contact and respiratory tract involvement, by wearing protective ensemble with self-contained breathing apparatus.

>100 to <700 ppm	SAR
700 ppm and greater	SCBA

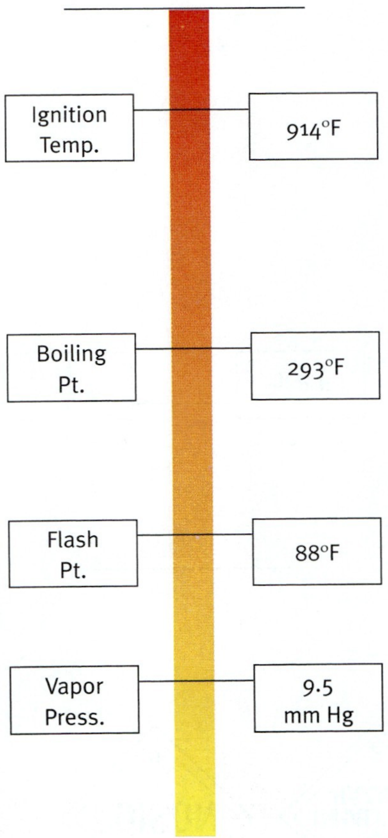

Ignition Temp. — 914°F

Boiling Pt. — 293°F

Flash Pt. — 88°F

Vapor Press. — 9.5 mm Hg

TOXICOLOGY

LEL	11,000 ppm	1.1%
LEL 10%	1,100 ppm	0.11%
LC$_{50}$	600 ppm	0.06%
IDLH	700 ppm	0.07%
Odor	0.047 ppm	0.000047%
STEL	200 ppm	0.02%
PEL	100 ppm	0.01%
TLV-TWA	50 ppm	0.005%

DECONTAMINATION

Wash immediately with soap and water, removing contaminated clothing.

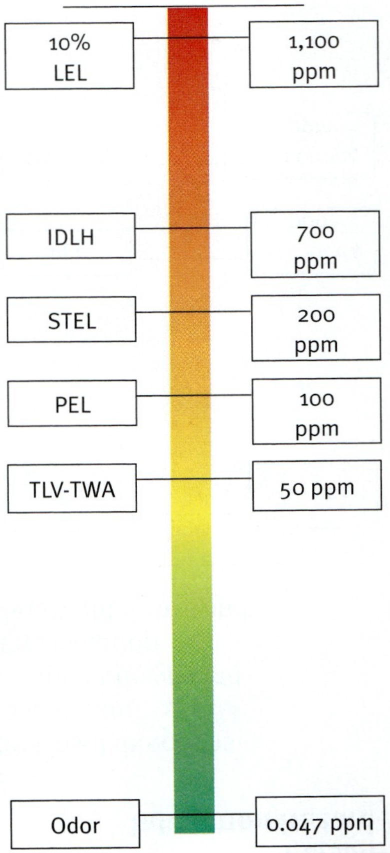

10% LEL — 1,100 ppm

IDLH — 700 ppm

STEL — 200 ppm

PEL — 100 ppm

TLV-TWA — 50 ppm

Odor — 0.047 ppm

SULFUR MUSTARD
$C_4H_8Cl_2S$

UN # 2810
CAS # 505-60-2
RTECS # WQ0900000
Military Designation: HD

GUIDE # 153

CBRNE Chemical
Poison
Inhalation Hazard

DOT-ERG Suggestions:
Fire Isolation: 800 meters in all directions
Spills: Small: 30 meters isolation
       0.1 mile day/night downwind
   Large: 60 meters isolation
       0.4–0.6 mile day/night downwind

CHEMISTRY

Vapor Pressure	0.072 mm Hg
Vapor Density	5.4
MW	159.08
UEL	
LEL	
Flash Point	
Ignition Temp.	
Specific Gravity	1.27
Solubility	Slightly soluble
Boiling Point	442.4°F
Ion Potential	? eV

TOXICOLOGY

LEL	
LEL 10%	
LC_{50}	1,500 mg-min/m³
IDLH	
Odor	Garlic/horseradish
STEL	
PEL	0.003 mg-min/m³
TLV-TWA	0.003 mg-min/m³

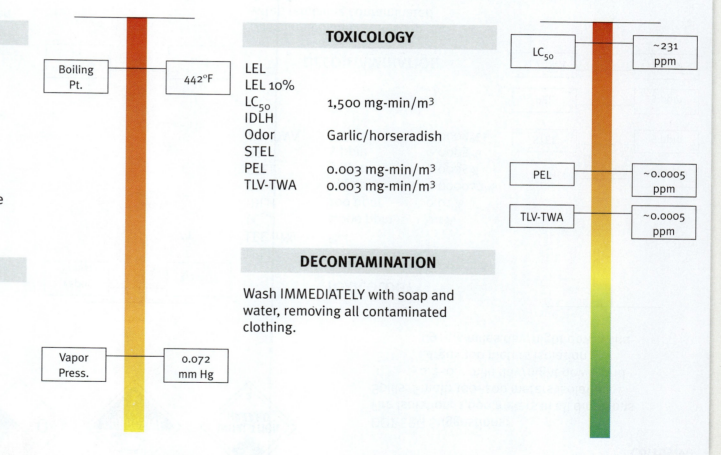

PPE

Wear the appropriate level of protection to prevent the possibility of skin contact and respiratory tract involvement, by wearing protective ensemble with self-contained breathing apparatus.

DECONTAMINATION

Wash IMMEDIATELY with soap and water, removing all contaminated clothing.

SULFUR DIOXIDE
SO₂

UN # 1079
CAS # 7446-09-5
RTECS # WS4550000

GUIDE # 125

Polymerization
Corrosive

DOT-ERG Suggestions:
Fire Isolation: 1,600 meters in all directions
Spills: Small: 100–200 meters isolation
        0.2–0.7 mile day/night downwind
      Large: 300 meters isolation
        1.9–4.5 miles day/night downwind

CHEMISTRY

Vapor Pressure	3.2 ATM
Vapor Density	2.263
MW	64.07
UEL	
LEL	
Flash Point	Nonflammable
Ignition Temp.	
Specific Gravity	1.59
Solubility	Slightly soluble
Boiling Point	14°F
Ion Potential	12.30 eV

PPE

Depending on the quantity of the spilled material, wear the appropriate level of protection to prevent the possibility of skin contact and respiratory tract involvement, by wearing protective ensemble with self-contained breathing apparatus.

>5 to <100 ppm	APR
100 ppm and greater	SCBA

Vapor Press. — 3.2 ATM

Boiling Pt. — 14°F

TOXICOLOGY

LEL		
LEL 10%	%	
LC₅₀	1,000 ppm	0.1%
IDLH	100 ppm	0.01%
Odor	0.47 ppm	0.000047%
STEL	5 ppm	0.0005%
PEL	5 ppm	0.0005%
TLV-TWA	2 ppm	0.0002%

DECONTAMINATION

Wash immediately with soap and water, removing contaminated clothing.

IDLH — 100 ppm
STEL — 5 ppm
PEL — 5 ppm
TLV-TWA — 2 ppm
Odor — 0.47 ppm

SULFURIC ACID
H₂SO₄

UN # 1830
1831 (fuming—Oleum)
1832 (spent)
CAS # 7664-93-9
RTECS # WS5600000

GUIDE # 137

Corrosive

DOT-ERG Suggestions:
Fire Isolation: 800 meters in all directions
Spills: Small: 50–100 meters isolation
Large: 800 meters isolation

CHEMISTRY

Vapor Pressure	Low to high
Vapor Density	3.4
MW	98.08
UEL	
LEL	
Flash Point	Nonflammable
Ignition Temp.	
Specific Gravity	1.84
Solubility	Miscible
Boiling Point	554°F
Ion Potential	? eV

PPE

Depending on the quantity of the spilled material, wear the appropriate level of protection to prevent the possibility of skin contact and respiratory tract involvement, by wearing protective ensemble with self-contained breathing apparatus.

>1 to <15 mg/m³   SAR
15 mg/m³ and greater   SCBA

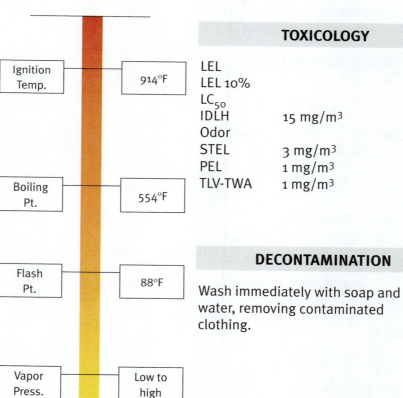

TOXICOLOGY

LEL	
LEL 10%	
LC₅₀	
IDLH	15 mg/m³
Odor	
STEL	3 mg/m³
PEL	1 mg/m³
TLV-TWA	1 mg/m³

DECONTAMINATION

Wash immediately with soap and water, removing contaminated clothing.

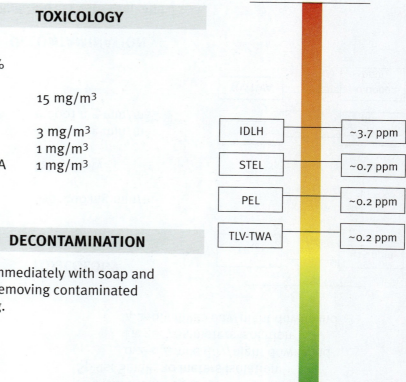

TABUN
$C_5H_{11}N_2O_2P$

UN # 2810
CAS # 77-81-6
RTECS # TB4550000
Military Designation: GA

POISON 6

INHALATION HAZARD 2

GUIDE # 153

CBRNE Chemical Poison Inhalation Hazard

DOT-ERG Suggestions:
Fire Isolation: 800 meters in all directions
Spills: Small: 30 meters isolation
      0.2–0.4 mile day/night downwind
     Large: 765 meters isolation
     4.2–6.5 miles day/night downwind

CHEMISTRY

Vapor Pressure	0.037 mm Hg
Vapor Density	5.63
MW	162.13
UEL	
LEL	
Flash Point	172°F
Ignition Temp.	
Specific Gravity	1.073
Solubility	Slightly soluble
Boiling Point	474.8°F
Ion Potential	? eV

PPE

Wear the appropriate level of protection to prevent the possibility of skin contact and respiratory tract involvement, by wearing protective ensemble with self-contained breathing apparatus.

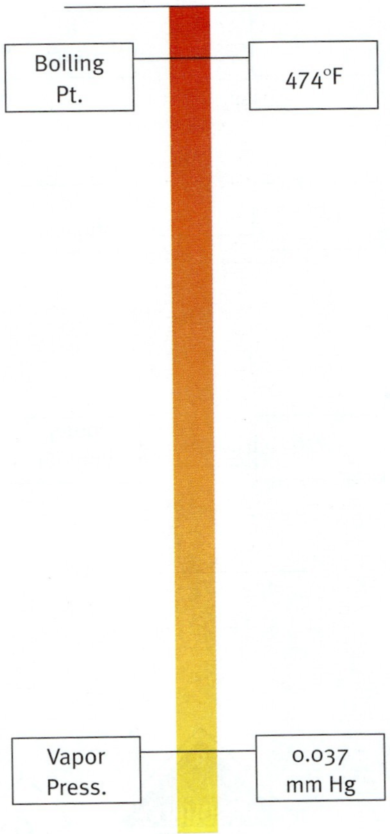

Boiling Pt. — 474°F

Vapor Press. — 0.037 mm Hg

TOXICOLOGY

LEL	
LEL 10%	
LC_{50}	100–200 mg-min/m³
IDLH	
Odor	Fairly fruity
STEL	
PEL	0.0001 mg-min/m³
TLV-TWA	0.0001 mg-min/m³

DECONTAMINATION

Wash IMMEDIATELY with soap and water, removing all contaminated clothing.

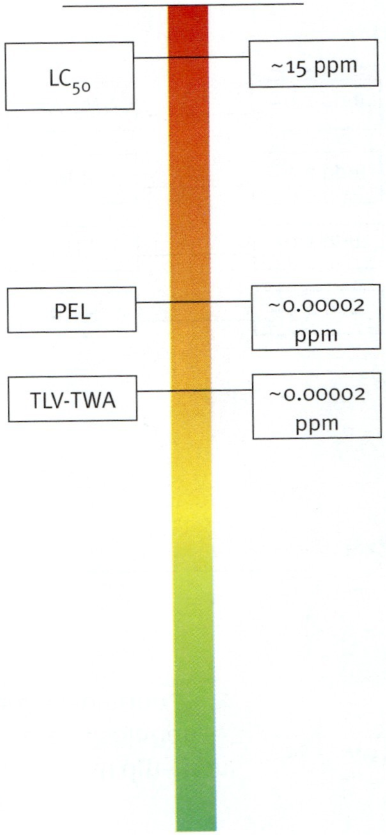

LC_{50} — ~15 ppm

PEL — ~0.00002 ppm

TLV-TWA — ~0.00002 ppm

TOLUENE
$C_6H_5CH_3$

UN # 1294
CAS # 108-88-3
RTECS # XS5250000

GUIDE # 130

Flammable

DOT-ERG Suggestions:
Fire Isolation: 800 meters in all directions
Spills: Small: 25–50 meters isolation
Large: 300 meters isolation

CHEMISTRY

Vapor Pressure	21 mm Hg
Vapor Density	3.14
MW	92.13
UEL	7.1%
LEL	1.2%
Flash Point	40°F
Ignition Temp.	896°F
Specific Gravity	0.8661
Solubility	Sparingly soluble
Boiling Point	232°F
Ion Potential	8.82 eV

PPE

Depending on the quantity of the spilled material, wear the appropriate level of protection to prevent the possibility of skin contact and respiratory tract involvement, by wearing protective ensemble with self-contained breathing apparatus.

>200 to <500 ppm   SAR
500 ppm and greater   SCBA

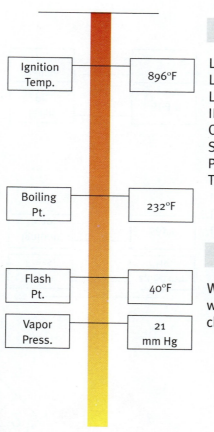

TOXICOLOGY

LEL	12,000 ppm	1.2%
LEL 10%	1,200 ppm	0.12%
LC_{50}	200 ppm	0.02%
IDLH	500 ppm	0.05%
Odor	2.14 ppm	0.000214%
STEL	300 ppm	0.03%
PEL	200 ppm	0.02%
TLV-TWA	100 ppm	0.01%

DECONTAMINATION

Wash immediately with soap and water, removing contaminated clothing.

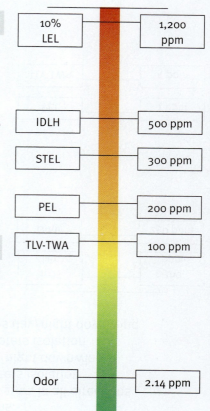

TOP 35 CHEMICALS / *Chemical Scan Sheets*

VINYL CHLORIDE
CH$_2$=CHCl

UN # 1086
CAS # 75-01-4
RTECS # KU9625000

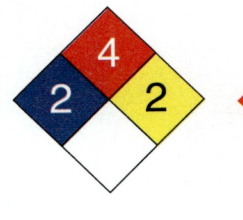

GUIDE # 130

**Polymerization
Flammable**

DOT-ERG Suggestions:
Fire Isolation: 800 meters in all directions
Spills: Small: 30 meters isolation
    0.1 mile day/night downwind
    Large: 60 meters isolation
    0.4–0.6 miles day/night downwind

CHEMISTRY

Vapor Pressure	3.3 ATM
Vapor Density	2.15
MW	62.5
UEL	33
LEL	3.6
Flash Point	−108°F
Ignition Temp.	881.6
Specific Gravity	0.912
Solubility	Slightly soluble
Boiling Point	8°F
Ion Potential	9.9 eV

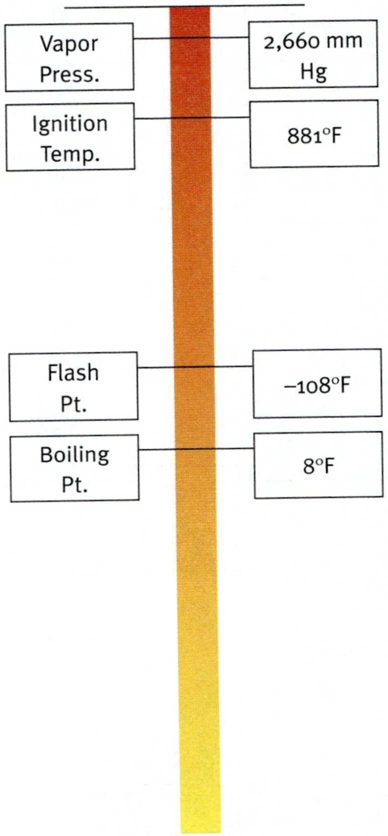

PPE

Depending on the quantity of the spilled material, wear the appropriate level of protection to prevent the possibility of skin contact and respiratory tract involvement, by wearing protective ensemble with self-contained breathing apparatus.

>1 to <1,000 ppm	SAR
1,000 ppm and greater	SCBA

Vapor Press.	2,660 mm Hg
Ignition Temp.	881°F
Flash Pt.	−108°F
Boiling Pt.	8°F

TOXICOLOGY

LEL	3,600 ppm
LEL 10%	360 ppm
LC$_{50}$	
IDLH	
Odor	250 ppm
STEL	
PEL	1 ppm
TLV-TWA	5 ppm

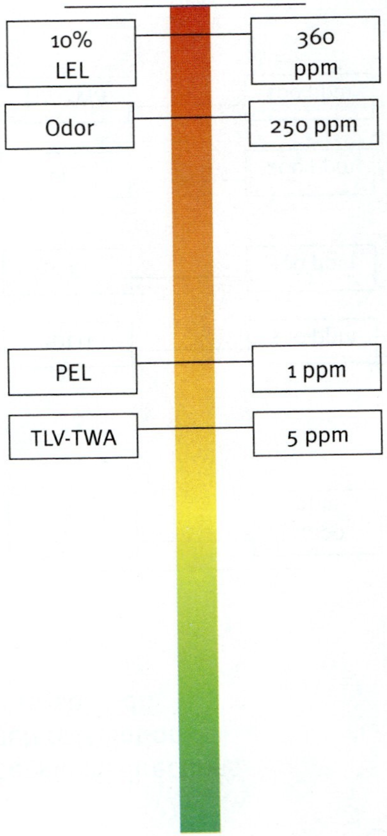

DECONTAMINATION

Wash immediately with soap and water, removing contaminated clothing.

10% LEL	360 ppm
Odor	250 ppm
PEL	1 ppm
TLV-TWA	5 ppm

VX (Ve, Vg, Vm)
$C_{11}H_{26}NO_2PS$

UN # 2810
CAS # 50782-69-9
RTECS # TB1090000
Military Designation: VX

GUIDE # 153

**CBRNE Chemical
Poison
Inhalation Hazard**

DOT-ERG Suggestions:
Fire Isolation: 800 meters in all directions
Spills: Small: 30 meters isolation
    0.2–0.4 mile day/night downwind
Large: 765 meters isolation
    4.2–6.5 miles day/night downwind

CHEMISTRY

Vapor Pressure	0.0007 mm Hg
Vapor Density	1.01
MW	267.38
UEL	
LEL	
Flash Point	
Ignition Temp.	
Specific Gravity	1.022
Solubility	Slightly soluble
Boiling Point	568.4°F
Ion Potential	? eV

TOXICOLOGY

LEL	
LEL 10%	
LC_{50}	5–15 mg-min/m³
IDLH	
Odor	Fairly fruity
STEL	
PEL	0.00001 mg-min/m³
TLV-TWA	0.00001 mg-min/m³

PPE

Wear the appropriate level of protection to prevent the possibility of skin contact and respiratory tract involvement, by wearing protective ensemble with self-contained breathing apparatus.

DECONTAMINATION

Wash IMMEDIATELY with soap and water, removing all contaminated clothing.

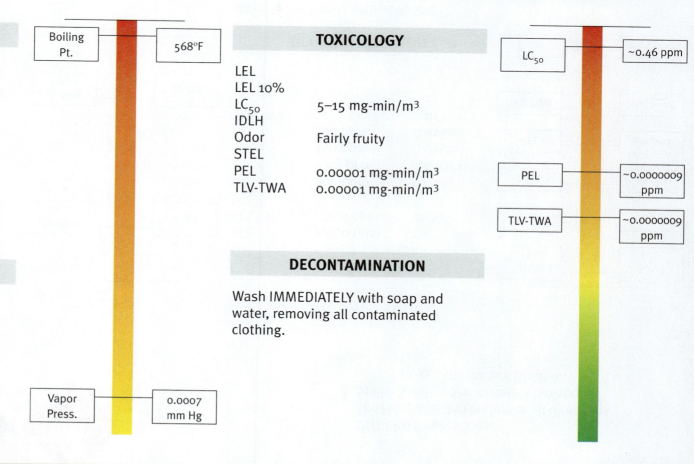

XYLENE
$C_6H_5(CH_3)_2$

UN # 1307
CAS # 95-47-6
　　　108-38-3
　　　106-42-3
RTECS # ZE2450000
　　　ZE2275000
　　　ZE2625000

GUIDE # 130　　　　　　　　　　　　　　　**Flammable**

DOT-ERG Suggestions:
Fire Isolation: 800 meters in all directions
Spills: Small: 50–100 meters isolation
　　　　Large: 300 meters isolation

CHEMISTRY

Vapor Pressure	9 mm Hg
Vapor Density	4.5
MW	106.17
UEL	7.5
LEL	1.7
Flash Point	79°F
Ignition Temp.	869°F
Specific Gravity	0.88
Solubility	Sparingly soluble
Boiling Point	280°F
Ion Potential	8.56 eV

PPE

Depending on the quantity of the spilled material, wear the appropriate level of protection to prevent the possibility of skin contact and respiratory tract involvement, by wearing protective ensemble with self-contained breathing apparatus.

TOXICOLOGY

LEL	7,000 ppm	1%
LEL 10%	1,700 ppm	0.1%
LC$_{50}$		
IDLH	900 ppm	0.09%
Odor	0.05 ppm	0.0001%
STEL	150 ppm	0.015%
PEL	100 ppm	0.01%
TLV-TWA	100 ppm	0.01%

DECONTAMINATION

Wash immediately with soap and water, removing contaminated clothing.

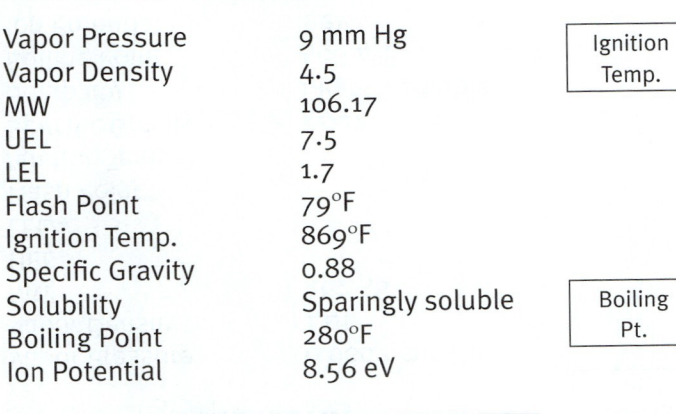

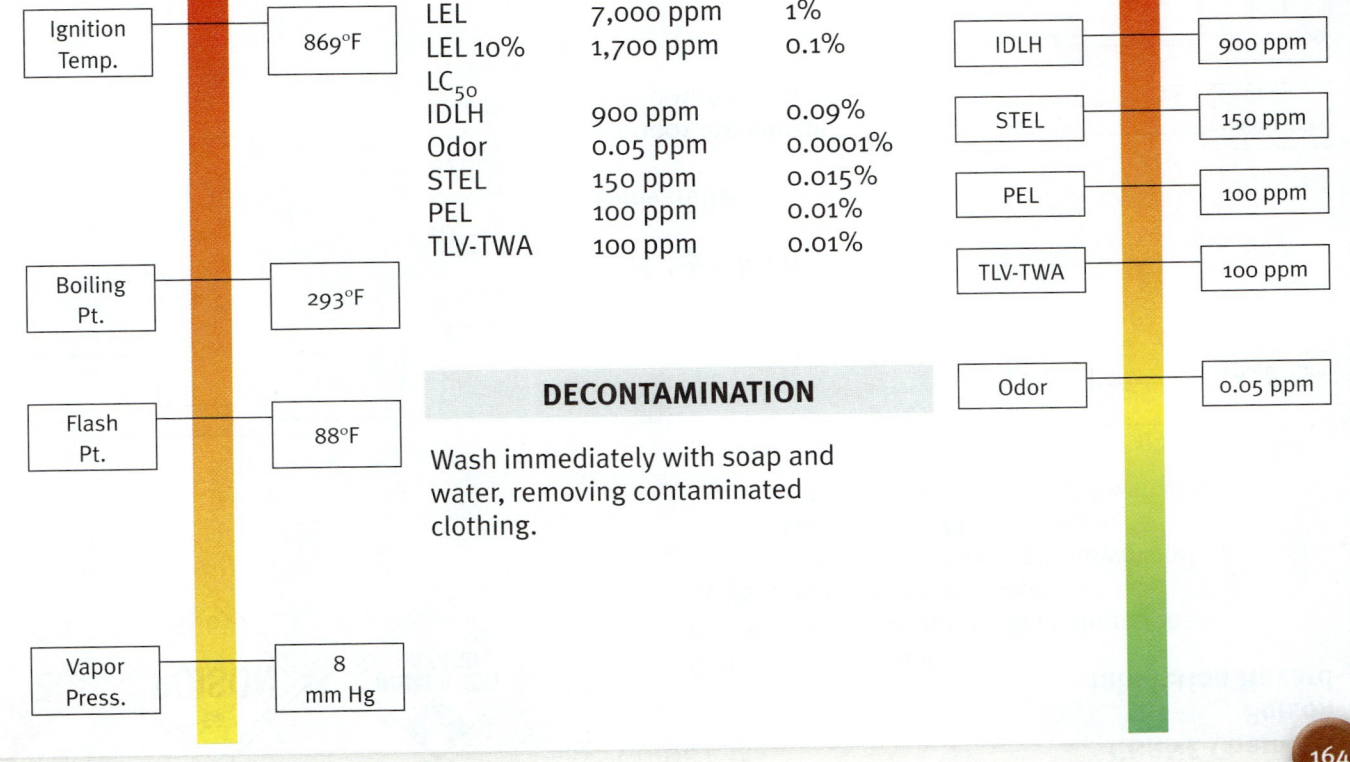

Ignition Temp. 869°F
Boiling Pt. 293°F
Flash Pt. 88°F
Vapor Press. 8 mm Hg

10% LEL 1,700 ppm
IDLH 900 ppm
STEL 150 ppm
PEL 100 ppm
TLV-TWA 100 ppm
Odor 0.05 ppm

164

CBRNE Tables

SECTION 6

- General Chemical Descriptions
- Chemical Agent Illicit Drug Precursors
- Biological Descriptions
- Isotope Descriptions
- Nuclear Descriptions
- Explosive Descriptions
- Estimations for Volume Using Formulae

CBRNE Chemical Agents — General Chemical Descriptions

Category	Chemical	Effects	Decontamination	Vapor Press. mm Hg	LC$_{50}$ mg/m^3
Nerve Agents	Tabun (GA) Sarin (GB) Soman (GD) GF Vx	Chemically similar to organophosphate pesticides, which exert an effect by inhibiting acetychlolinesterse. Readily absorbed by inhalation, dermal absorption, and ingestion. Rapid onset: seconds to minutes.	Soap and water wash, primary objective is to remove the substance from the individual rapidly. Removal of clothing may assist in this process. Rapid decon with water. Soap and water is preferred.	0.037 2.1 0.40 0.044 0.0007	400 100 50 — 10
Blister Agents	Nitrogen mustard (HN-1) (HN-2) (HN-3)	A vesicant, which causes cellular changes through alklation. Clinical effects are delayed for several hours. Readily absorbed by inhalation, dermal contact, and ingestion.	Soap and water wash—primary objective is to remove the substance rapidly. Removal of clothing may assist in this process. Rapid decon with water. Soap and water wash is preferred.	0.011–0.29	1,500–3,000
	Sulfur mustard (H) (HD)	A vesicant that causes cellular changes through alkylation. Clinical effects are delayed 1–24 hours. Readily absorbed by inhalation, dermal contact, and ingestion.	Soap and water wash—primary objective is to remove the substance rapidly. Removal of clothing may assist in this process. Rapid decon with water. Soap and water wash is preferred.	0.072	1,500
	Lewisite (L)	The airway is the primary target for this chemical. Causes pain on contact with extreme irritating effects, which causes cellular damage, progressing into a variety of complications.	Soap and water wash—primary objective is to remove the substance rapidly. Removal of clothing may assist in this process. Rapid decon with water. Soap and water wash is preferred.	0.4	1,400
	Phosgene oxime (CX)	Causes pain upon contact, which is followed by blanching, as does the other visicants. Pain on contact with extreme irritating effects, which causes cellular damage, progressing into a variety of complications.	Soap and water wash—primary objective is to remove the substance rapidly. Removal of clothing may assist in this process. Rapid decon with water. Soap and water wash is preferred.	11–13	3,200
Blood Agents	Cyanogen chloride (CK) Hydrogen cyanide (AC)	Cyanides cause progressive tissue hypoxia. Combination with cytochrome oxidase at the cellular level. Biting odor, which is sometimes recognized as bitter almonds smell.	Soap and water wash—primary objective is to remove the substance rapidly. Removal of clothing may assist in this process.	1,000 612	11,000 2,000
Choking Agents	Chlorine (CL)	A very corrosive and destructive chemical to tissues. Immediate onset with damage dependent on the concentration and length of contact.	Soap and water wash—primary objective is to remove the substance rapidly. Removal of clothing may assist in this process.	5 ATM	500 ppm
	Phosgene (CG)	Effects of exposure can take up to 24 hours before victim identifies a health issue. Severe irritation and aggressive destruction of the respiratory tree.	Soap and water wash—primary objective is to remove the substance rapidly. Removal of clothing may assist in this process.	1,173	3,200

Chemical Agent Illicit Drug Precursors

Chemical Precursors	GA	GB	GD	Vx	HN	HD	L	CK	AC	Amphetamine	GHB	Ecstacy	Meth	Cathinone	PCP
2-Chloroethanol					X	X									
Ammonium bifluoride		X	X												
Ammonium formate										X					
Anhydrous ammonia													X		
Arsenic trichloride								X							
Benzaldehyde										X					
Bromobenzene															X
Butylamine										X		X			
Chloroform													X		
Chromium trioxide														X	
Cyanide salts															X
Diethyl ethylphosphonate		X													
Diethyl methylphosphonite				X											
Diethyl N,N-dimethyl phosphoramidate	X														
Diethylaminoethanol				X											
Diethylphosphite		X	X	X											
Diisopropylamine				X											
Dimethyl ethylphosphonate		X													
Dimethyl methylphosphonate		X	X												
Dimethylamine	X														
Dimethylamine HCl	X														
Dimethylphosphite		X	X												
Ethanol											X		X		
Ethylphosphonous dichloride															
Ethylphosphonous difluoride		X		X											
Ethylphosphonyl dichloride		X													
Ethylphosphonyl difluoride		X													

Chemical Precursors	GA	GB	GD	Vx	HN	HD	L	CK	AC	Amphetamine	GHB	Ecstacy	Meth	Cathinone	PCP
Formamide										X					
Gamma butyrolactone											X				
Hydrochloric acid												X			
Hydrogen bromide												X			
Hydrogen chloride													X		
Hydrogen fluoride		X	X												
Hypophosphorus acid													X		
Iodine													X		
Lithium													X		
Magnesium															
Mercuric chloride													X		
Methylamine												X	X		
Methylphosphonous dichloride				X											
Methylphosphonous difluoride		X	X	X											
Methylphosphonyl dichloride		X	X												
Methylphosphonyl difluoride		X	X												
N,N-diisopropyl-2-aminoethyl chloride hydrochloride				X											
N,N-diisopropyl-aminoethanathiol				X											
Nitroethane										X			X		
Phenyl-2-propanone										X					
Phosphorus													X		
Phosphorous oxychloride	X														
Phosphorous pentachloride	X														
Phosphorous pentasulfide				X											
Phosphorous trichloride	X	X	X												
Pinacolone			X												

Chemical Precursors	GA	GB	GD	Vx	HN	HD	L	CK	AC	Amphetamine	GHB	Ecstacy	Meth	Cathinone	PCP
Pinacolyl alcohol			X												
Piperonal												X			
Potassium bifluoride		X	X												
Potassium cyanide	X								X						
Potassium fluoride		X	X												
Sodium													X		
Sodium bifluoride		X	X												
Sodium cyanide	X							X	X						
Sodium cyanoborohydride												X			
Sodium dichromate														X	
Sodium fluoride		X	X												
Sodium hydroxide											X				
Sodium sulfide						X									
Sulfur dichloride						X									
Sulfur monochloride						X									
Sulfur sulfate									X						
Sulfuric acid													X	X	
Thiodiglycol						X									
Thionyl chloride		X	X		X	X									
Toluene														X	
Triethanol amine					X										
Triethanol amine hydrochloride					X										
Triethyl phosphite				X											
Trimethyl phosphite		X	X												

CBRNE CHEMICAL TABLES/*Chemical Agent Illicit Drug Precursors*

Biological Descriptions

Bacteria	Anthrax	Cholera	Plague	Tularemia
General Information	Sometimes described as "black as coal," splenic fever, or woolsorter's disease.	An intestinal disease that is caused by bacteria found in contaminated food/water due to poor sewage precautions	A bacteria that exists in several forms; the vectors are fleas or contaminated animal tissue. The plague is characterized by bubonic-swollen lymph nodes.	A bacteria that is very infectious. No person-to-person transmission, but can survive at low temperatures. The disease is characterized by fever, localized skin or mucous membrane ulceration, and occasional pneumonia.
	A naturally occurring zoonotic that the organism can sporolate	It is the toxin that is released by bacteria that causes the syndrome. Is not transmitted through person-to-person contact.	Septicemic events are its normal course.	Naturally occurring spread by rodents that have the tick, rodent droppings.
	No person-to-person transmission. Inhalation or dermal contact with spores.	Weapon—water contamination.	Weapon—aerosolized	Weapon—aerosolized
	Weapon—aersoilized spores	Chlorine/filter	100% mortality if not treated.	
		Mostly incapacitating		
Clinical Presentation	Cutaneous	Bacterium attaches to the tissue of the small intestine releasing a toxin that causes oversecretion of fluid	Enlarged lymph nodes	Skin lesions or on the mucous membranes (including the conjunctiva). Enlarged or smaller than normal lymph nodes. Pneumonia and associated symptoms.
	Redden and elevated area (papule), which forms a vesicle (fills with fluid) and then forms a scab—black eschar	Watery stool 5-10 L/day	Infectious disease	Incubation 1–10 days
	Inhalation injury is the same within the lungs	Rice water	Incubation	
	Incubation 1–6 days	Incubation 1–5 days	Bubonic 2–10 days	
			Pneumonic 2–5 days	
Signs and Symptoms	Fever	Anorexia	Fever/chills	Fever/chills
	Malaise	Vomiting	Malaise	Headache
	Mild chest discomfort	Diarrhea	Diffuse pain	Cough
	Dry cough	Dehydration	Cough—pink sputum	Myalgias
	Anthrax eclipse	Electrolyte loss	Dyspena	Chest pain
	Widening mediastinum	Shock	Cyanosis	Vomiting
	Hemorrhagic mediastinitis	Renal failure	Toxemia	Abdominal pain
	Death in 24–36 hours	Death in hours to days	Septicemic—CNS	Diarrhea
				Back pain and/or stiff neck

Virus	Smallpox	Encephalitis	Viral Hemorrhagic Fever	Hantavirus
General Information	Eradicated in 1980	Various forms	Ebola Marburg Filoviridae family	A virus that is normally found in rodents; limited information on person-to-person transmission. A field mouse is the host in which the organism sequesters. Events of infection have been due to the drying of fecal material and aersolizing during cleaning of an area.
	Last vaccination 1980	Disease in horses and mules	Lassa, Argintine, Bolivian Arenaviridae family	
	Early stages resemble chickenpox	Vector mosquito	Hantavirus Crimean-Congo Bunyaviridae family	
	Patient is highly contagious until scabs form	Aerosolized weapon	Yellow fever and dengue Flaviviridae family	
	Only two known repositories: CDC and Vector	Natural outbreak will have animals infected first and outside normal areas of South and Central America, Mexico, Trinidad, and Florida		
Clinical Presentation	Rash, which is followed by red, raised papules that form pustular vesicles. The vesicles that are common on the extremities, and face in chickenpox concentrate on the chest and abdomen.	Inflammation of the meninges	Highly lethal	Rapid onset of fever with vomiting, headache, and severe abdominal pain lasting 3–10 days. Hemorrhagic fever syndrome starts from days 2–3 to days 6–7. Approximately 50% mortality rate.
	Incubation 12 days	Susceptibility 90–100% with fatality less than 1%	Highly contagious	
		Incubation 1–5 days	Incurable	
			No vaccines	
Signs and Symptoms	Acute fever	Sudden onset	Rapid on set of fever	Rapid onset of fever, vomiting, headache, severe abdominal pain, anorexia, back pain, and severe prostration
	Malaise	Profound onset	Malaise	
	Vomiting	Headache	Easy bleeding	
	Headache	Fever	Hypotension	
	3–30% mortality	Malaise	Shock	
	Quarantine all exposed for 17 days	Photophobia	Mortality 5–90%	
		Malaise	Diagnostic testing CDC	
		Photophobia		

CBRNE CHEMICAL TABLES / *Biological Descriptions*

Toxins	Staphylococcal Enterotoxin B	Ricin	Botulinum	Trichothecene Mycotoxin
General Information	Toxin produced by *Staphylooccus aureus* endotoxin affects the intestines. SEB food poisoning from improperly handled or cooked food.	A protein with tox level 6,000 times greater than CN	One of the most potent toxins; LD_{50} is 0.001 μg/kg or, for a 220-lb man, 0.1 μg	Naturally produced by fungi. Inhibits protein synthesis by destroying the integrity of the membrane—targets the most rapidly growing cells, skin, mucous membranes, and bone marrow. T2 can enter all routes
	Considered as an incapacitating agent in up to 80% of cases, with down time of two weeks or more. Could be used to contaminated low-volume water supply.	Interrupts protein synthesis, altering RNA	15,000 times more toxic than VX. The toxin is resistive to the digestive acids.	
			Found in poorly handled food	
			Weapon—aerosolized	
Clinical Presentation	Clinical picture is dependent on mode of exposure.	Dependent on route of entry	Binds to the membrane of the pre-synaptic cholinergic neurons and prevents ACh release.	Burning and itching , reddened skin, burning in the nose and throat, sneezing, burning of the eyes, and conjunctivitis. The burning tissue advances toward blacked necrosed tissue.
	Normal food is clustered	Inhalation—nonspecific pulmonary	Classified as a neurotoxin	Aerosolized—yellow rain
	Proliferation of T-cells reaction similar to organ rejection	Ingestion—gastro hemorrhage	Permeability of the vascular system	
		Intramuscular local lymph necrosis	Incubation dependent on type	
Signs and Symptoms	Onset 3–12 hours	Onset 3 hours but commonly 8 –12 hours Inhalation—necrosis of tissue, bronchitis, interstitial pneumonia, and PE	Onset 8–36 hours	Onset a few minutes to a few hours
	Based on portal of entry	Fever, cough, chest discomfort, weakness Ingestion—hepatic and splenic necrosis, present as GI bleed	Dry mouth, Difficulty speaking and swallowing	GI nausea, vomiting, diarrhea, and abdominal pain
	Nonspecific flulike		Blurred vision	
	Rapid fever 103–106°F		Photophobia dilated pupils	
	Headache		Flaccid muscles	
	General weakness		Bilateral and symmetrical descending weakness and paralysis	
	Nonproductive cough			
	Severe: substernal chest pain			
	Dspena PE			
	Food poison—epidemic			

Isotope Descriptions

Radionuclide	Symbol	Radiation Type	R/hr per g at 30 cm	Radionuclide	Symbol	Radiation Type	R/hr per g at 30 cm
Americium 241	241Am	Alpha, gamma	0.58	Curium 242	242Cm	Alpha, gamma	
Americium 243	243Am	Alpha, gamma	0.23	Curium 243	243Cm	Alpha, gamma	
Antimony 122	122Sb	Beta, gamma	N/A	Curium 244	244Cm	Alpha, gamma	2,650
Antimony 124	124Sb	Beta, gamma	N/A	Dysprosium 159	159Dy	Gamma	
Antimony 125	125Sb	Beta, gamma	N/A	Erbium 169	169Er	Beta, gamma	
Argon 37	37Ar	Gamma		Europium 152	152Eu	Beta, gamma	
Arsenic 74	74As	Beta, gamma		Europium 154	154Eu	Beta, gamma	
Arsenic 76	76As	Beta, gamma		Europium 155	155Eu	Beta, gamma	
Arsenic 77	77As	Beta, gamma		Fluorine 18	18F	Beta, gamma	735,000,000
Barium 131	131Ba	Gamma		Gadolinium 153	153Gd	Gamma	
Barium 133	133Ba	Gamma		Gallium 68	68Ga	Beta, gamma	
Barium 137	137Ba	Gamma	2,390,000,000	Gallium 72	72Ga	Beta, gamma	
Barium 140	140Ba	Beta, gamma	132,000	Germanium 71	71Ge	Gamma	
Beryllium 7	7Be	Gamma	133,000	Gold 195	195Au	Gamma	
Bismuth 207	207Bi	Gamma	N/A	Gold 198	198Au	Beta, gamma	
Bismuth 210	210Bi	Alpha, beta, gamma	N/A	Gold 199	199Au	Beta, gamma	
Bromine 82	82Br	Beta, gamma		Hafnium 181	181Hf	Beta, gamma	
Cadmium 115	115Cd	Beta, gamma		Holmium 166	166Ho	Beta, gamma	
Calcium 45	45Ca	Beta		Hydrogen 3	3H	Beta	N/A
Calcium 47	47Ca	Beta, gamma		Indium 113	113In	Gamma	
Californium 252	252Cf	Alpha, gamma	248	Indium 114	114In	Beta, gamma	
Carbon 14	14C	Beta	N/A	Iodine 125	125I	Beta, gamma	53,100
Cerium 141	141Ce	Beta, gamma		Iodine 129	129I	Beta, gamma	0.00247
Cerium 144	144Ce	Beta, gamma		Iodine 130	130I	Beta, gamma	
Cesium 131	131Cs	Gamma		Iodine 131	131I	Beta, gamma	389,000
Cesium 134	134Cs	Beta, gamma		Iridium 192	192Ir	Beta, gamma	60,400
Cesium 137	137Cs	Beta, gamma	N/A	Iridium 194	194Ir	Beta, gamma	
Chlorine 36	36Cl	Beta, gamma	N/A	Iron 55	55Fe	Gamma	N/A
Chromium 51	51Cr	Gamma	14,800	Iron 59	59Fe	Beta, gamma	365,000
Cobalt 57	57Co	Gamma	14,200	Krypton 85	85Kr	Beta, gamma	7.85
Cobalt 58	58Co	Gamma	216,000	Lanthanum 140	140La	Beta, gamma	6,900,000
Cobalt 60	60Co	Beta, gamma	17,200	Lead 210	210Pb	Alpha, beta, gamma	213
Copper 64	64Cu	Beta, gamma		Lutetium 177	177Lu	Beta, gamma	

Radionuclide	Symbol	Radiation Type	R/hr per g at 30 cm	Radionuclide	Symbol	Radiation Type	R/hr per g at 30 cm
Magnesium 28	28Mg	Beta, gamma		Silver 110	110Ag	Beta, gamma	
Manganese 54	54Mn	Gamma	43,900	Silver 111	111Ag	Beta, gamma	
Mercury 197	197Hg	Gamma		Sodium 22	22Na	Beta, gamma	92,700
Mercury 203	203Hg	Beta, gamma		Sodium 24	24Na	Beta, gamma	179,000,000
Molybdenum 99	99Mo	Beta, gamma	600,000	Strontium 85	85Sr	Gamma	
Neodymium 147	147Nd	Beta, gamma		Strontium 87	87Sr	Gamma	
Neptunium 237	237Np	Alpha, gamma	0.00362	Strontium 89	89Sr	Beta, gamma	26.1
Neptunium 239	239Np	Beta, gamma	1,320,000	Strontium 90	90Sr	Beta	
Nickel 63	63Ni	Beta		Sulfur 35	35S	Beta	
Niobium 95	95Nb	Beta, gamma		Tantalum 182	182Ta	Beta, gamma	
Palladium 103	103Pd	Gamma		Technetium 99	99Tc	Beta, gamma	7,160,000
Palladium 109	109Pd	Beta, gamma		Tellurium 132	132Te	Beta, gamma	
Phosphorus 32	32P	Beta	N/A	Terbium 160	160Tb	Beta, gamma	
Plutonium 238	238Pu	Alpha, gamma	14.9	Thallium 204	204Tl	Beta, gamma	5.75
Plutonium 239	239Pu	Alpha, gamma	0.0205	Thorium 230	230Th	Alpha, gamma	0.0157
Polonium 210	210Po	Alpha, gamma		Thorium 232	232Th	Alpha, gamma	0.0000000837
Potassium 42	42K	Beta, gamma	18,300,000	Thorium 232N	232Th	Alpha, beta, gamma	
Praseodymium 142	142Pr	Beta, gamma		Thulium 170	170Tm	Beta, gamma	
Praseodymium 143	143Pr	Beta		Tin 113	113Sn	Gamma	
Praseodymium 144	144Pr	Beta, gamma		Tin 119	119Sn	Gamma	
Promethium 147	147Pm	Beta		Titanium 44	44Ti	Gamma	
Promethium 149	149Pm	Beta, gamma		Tungsten 185	185W	Beta	
Protactinium 233	233Pa	Beta, gamma		Tungsten 187	987W	Beta, gamma	
Protactinium 234	234Pa	Beta, gamma	15,100,000,000	Uranium 235	235U	Alpha, gamma	813
Radium 224	224Ra	Alpha, gamma	19,100	Uranium 238	238U	Alpha, gamma	0.000000242
Radium 226	226Ra	Alpha, gamma	0.129	Uranium 238N	238U	Alpha, beta, gamma	
Rhenium 186	186Re	Beta, gamma		Xenon 133	133Xe	Beta, gamma	
Rhodium 106	106Rh	Beta, gamma		Ytterbium 169	169Yb	Gamma	
Rubidium 86	86Rb	Beta, gamma	432,000,000	Yttrium 90	90Y	Beta	
Ruthenium 103	103Ru	Beta, gamma		Yttrium 91	91Y	Beta, gamma	
Ruthenium 106	106Ru	Beta		Zinc 65	65Zn	Beta, gamma	30,200
Ruthenium 97	97Ru	Gamma		Zinc 69	69Zn	Beta	
Samarium 151	151Sm	Beta, gamma		Zirconium 95	95Zr	Beta, gamma	111,000
Samarium 153	153Sm	Beta, gamma					

Adapted from NRC

Nuclear Descriptions

Dose	Signs and symptoms
0–25 RAD	Subclinical range: Nausea and fatigue if any symptomology, no detectable effects
25–100 RAD	A percentage of the population will present with nausea and vomiting, anorexia, and fatigue Decreased red and white blood cell counts; platelets and lymphocytes may be affected Bone marrow damage noted
100–300 RAD	Mild to severe nausea and vomiting, anorexia and fatigue, with infection concerns Severe damage to the hematological system
300–600 RAD	Severe nausea and vomiting, anorexia and fatigue, with infection concerns Severe damage to the hematological system, massive bleeding with infection, and diarrhea Fatality rate starts to approach 50% of the population affected
600 + RAD	Central nervous system fails, along with hematological and immune systems—death probable

Acute Radiation Syndrome

Phase	System Failure	Subclinical Range		Sublethal Range		Lethal Range	
		0–100 RAD	100–200 RAD	200–600 RAD	600–800 RAD	800–3,000 RAD	3,000 RAD +
Prodromal phase	Lympocyte count		Minimal	<1,000 @ 24 hrs	<500 @ 24 hrs	Rapid decrease within hours	
	Nausea, vomiting		5 – 50%	50 – 100%	75 – 100%	90 – 100%	100%
	CNS function			Cognitive impairment 6–20 hrs	Cognitive impairment 24 hrs+	Rapid incapacitation	
	Time of onset		3–6 hrs	2–4 hrs	1–2 hrs	< 1 hr	Within minutes
Latent phase	Absence of symptoms	2 weeks +	7–15 days	0–7 days	0–2 days	Rapid incapciation no symptoms	
Acute radiation Lethargy Illness	Signs and symptoms		Moderate leukopenia	Severe leukopenia, hemorrhaging, hair loss		Diarrhea Fever Electrolyte balance disrupted	Convulsions Tremors
	Time of onset		2 weeks +	2 days–2 weeks 4–6 highest potential for effective medical care		1–3 days	
Mortality			Minimal	Low with aggressive medical care	High	Neurological symptoms indicate lethal dose	

Adapted from NRC

Dose Equivalent Calculations

1 Roentgen (R)	2.58×10^{-4} coulomb/kg
1 Roentgen/hr (R/hr)	1×10^{-4} 13 amperes/cm^3
1 Radiation Absorbed Dose (RAD)	100 ergs/g
1 Becquerel (Bq)	2.27×10^{-11} Ci
1 Gray (Gy)	100 RAD
1 Sievert (Sv)	100 REM
1 Roentgen Equivalent Man (REM)	0.01 Sv
1 Radiation Absorbed Dose (RAD)	0.01 Gy
1 Roentgen (R)	2.58×10^{-4} C/kg
1 Curie (Ci)	3.7×10^{-10} Bq

Dose Limit (whole body)	Emergencycy Action Dose Guidelines Activity Performed
5 rem	All activities
10 rem	Protecting major property
25 rem	Lifesaving or protection of large populations
>25 rem	Lifesaving or protection of large populations, only by volunteeers who understand the risks.

Inverse Square Law – Exposure-Distance Calculation

E estimated = E meter reading (D meter distance)^2 / (D estimated distance)^2

CBRNE Explosive Agents — Explosive Descriptions — ERG—372

Improvised Explosive Device (IED)

Threat Characteristics	TNT Equivalent	Inside Evacuation	Outside Evacuation
Pipe bomb	5 lb	70 ft	850 ft
Suicide belt/vest	10–20 lb	90–100 ft	1,080–1,360 ft
Breifcase/suitcase	50 lb	150	1,850 ft
Compact sedan	500 lb	320 ft	1,500 ft
Sedan	1,000 lb	400 ft	1,750 ft
Passenger van	4,000 lb	640 ft	2,700 ft
Delivery truck	10,000 lb	860 ft	3,750 ft
Moving van	30,000 lb	1,250 ft	6,500 ft
Semi-trailer	60,000 lb	1,575 ft	7,000 ft

Improvised Explosive Device (IED) Involving LPG

Threat Characteristics	TNT Equivalent	Inside Evacuation	Outside Evacuation
Small BBQ tank	20 lb–5 gallons	40 ft	160 ft
Large forklift tank	100 lb–25 gallons	70 ft	276 ft
Commerical tank	2,000 lb–500 gallons	184 ft	736 ft
Small truck (bobtail)	8,000 lb–2,000 gallons	292 ft	1,200 ft
Semi-tanker	40,000 lb–10,000 gallons	500 ft	2,000 ft

Adapted from TSWG

Blast Wave Dynamics

Pressure in PSI	Potential Injuries	Structural Effects
0.5–3	Rupture of eardrums, victim placed off balance	Facades fail, glass shatters
5–6	Internal organs rupture, shatter, spald	Steel structure fails, containers collapse, cinderblock shatters, utility poles fail
15	Multisystem trauma	Typical construction failure
30	Lungs collapse, internal organs fail	Reinforced construction fails
100	Fatal	Structural failure

Developed by ATF with technical assistance from U.S. Corp of Engineers support by Technical Support Working Group

Estimations for Volume Using Formulae

Intermediate Bulk Containers

Using formulae for a cube to estimate volume and hence gallons contained

Volume = A^3
Where a = length $\times$ width $\times$ height
Answer is in cubic ft, multiply by 7.48 (7.5) for conversion to gallons

If your IBC is 48 inches by 48 inches, and 48 inches height
48/12 = 4 ft
4 ft $\times$ 4 ft $\times$ 4 ft = 64 cubic ft $\times$ 7.5 = **480 gallons**

Nonbulk - Drums

Using formulae for a cylinder to estimate volume and hence gallons contained

Volume = area of the base $\times$ height
Where area of the base = πr^2 = 3.1416r^2
Answer is in cubic ft, multiply by 7.48 (7.5) for conversion to gallons

Standard 55-gallon drum is 23 inches in diameter and 35 inches in height
23 inches /12 = 1.9167 ft and 35 inches/12 = 2.9167 ft
r = radius of the diameter 1.9167 or 1.9167/2 = 0.958
3.1416 $\times$ (0.958)^3 $\times$ 2.9167 = 8.409 multiply by 7.5 for gallons = **63.07 gallons**

Rail Tank Car (or Cargo Tank Car)

Using formulae for a cylinder to estimate volume and hence gallons contained

Volume = area of the base $\times$ height
Diameter 15 ft, length 68 ft
Radius = 7.5 ft
3.1416 $\times$ (7.5)^3 $\times$ 68 = 1201.66 multiply by 7.5 for gallons = **90,125 gallons**

Horizontal Spherical Tank

Volume of a sphere = 4/3 πr^2, answer in cubic feet, multiply by 7.5 for conversion to gallons